ENVIRONMENTAL RESEARCH ADVANCES

MOLECULAR EVOLUTION OF AQUATIC ORGANISMS' BIOMARKERS

ENVIRONMENTAL RESEARCH ADVANCES

Additional books and e-books in this series can be found on Nova's website under the Series tab.

ENVIRONMENTAL RESEARCH ADVANCES

MOLECULAR EVOLUTION OF AQUATIC ORGANISMS' BIOMARKERS

RIGERS BAKIU

nova
science publishers
New York

NOTICE TO THE READER

Library of Congress Cataloging-in-Publication Data

ISBN: 978-1-53614-846-6

Published by Nova Science Publishers, Inc. † New York

To all my big family and my friends for their great support during the time of writing this book

CONTENTS

PREFACE

Ensuring the health of aquatic ecosystems and identifying species at risk from the detrimental effects of environmental contaminants can be facilitated by integrating analytical chemical analysis with carefully selected biological endpoints measured in tissues of species of concern. These biological endpoints include molecular, biochemical and physiological markers (i.e., biomarkers) that when integrated, can clarify issues of contaminant bioavailability, bioaccumulation and ecological effects while enabling a better understanding of the effects of non-chemical stressors. In the case of contaminant stressors, an understanding of chemical modes of toxicity can be incorporated with diagnostic markers of aquatic animal physiology to help understand the health status of aquatic organisms in the field. New approaches in functional genomics and bioinformatics can help discriminate individual chemicals, or groups of chemicals among complex mixtures that may contribute to adverse biological effects. By using these recent methodologies, it could be useful to put light on the molecular evolution of the biomarkers, which role and functionality can be better understood based on exploring the relative evolution pathways in several aquatic organisms.

Chapter 1

INTRODUCTION

At the beginning of the 1990s, ecotoxicologists became interested in the biomarker concept, which stimulated important debate, for instance, at the 2nd European Conference on Ecotoxicology organized by the Society of Ecotoxicology and Environmental Safety (SECOTOX) in Amsterdam in 1992.

In 1994, Depledge proposed a definition that is still authoritative today: "A biochemical, cellular, physiological or behavioural variation that can be measured in tissue or body fluid samples or at the level of whole organisms that provides evidence of exposure to and/or effects of, one or more chemical pollutants."

In the United States, the Clean Water Act is the primary federal law governing water pollution. The United States Environmental Protection Agency has developed a strategy to improve monitoring and assessment of environmental risk in aquatic ecosystems (at local, state, regional, and national scales), because of its statutory responsibilities. In this framework, the Environmental Monitoring and Assessment Program (EMAP) has substantially advanced the scientific basis for monitoring the condition of aquatic ecosystems. The EMAP strategy includes physicochemical indicators in sediments and the water column and, for biological indicators, mainly responses at the level of the community.

The Water Framework Directive (WFD) promulgated by the European Parliament and Council is the chosen way forward to maintain or improve the quality of European aquatic environments. In this aim, it is necessary to attain a good status of these waters. This good status is based on both the chemical and the ecological status of the water masses. The chemical status is considered "good" when the concentrations of chemicals in the medium are below the limits defined in European Comission's regulations. The characterization of the ecological status of water masses is mainly based on the composition and abundance of certain plant and animal taxa. The failure of the WFD to recognize a role for biomarkers in this context is regrettable as is their limited use in the EMAP strategy.

Anyway, by neglecting biomarkers, both regulatory bodies ignore a category of biological tools well known to be precocious and sensitive indicators of the degradation of organism health. Effects at the community level allow an ecotoxicological assessment after severe environmental degradation has already occurred, thus leading to expensive remediation processes, whereas biomarkers have an interesting potential as predictive tools usable much earlier in any environmental degradation process.

Ecological analyses recommended in the EMAP or the WFD are useful to describe differences between sites, differently impacted by anthropogenic pressure, or to reveal temporal changes, when historical records are available at moment of analyses performing. However, ecological approaches are of no help in determining the origin of such changes, whereas so-called "specific" biomarkers can contribute to answering this type of question.

Chemical data needed to fulfill the requirements of the WFD or the EMAP strategy may be useful to predict the potential effects on living organisms but only if the dose–effect relationship is well established. Predicted No Effect Concentrations can be derived from laboratory toxicological tests, but the main limit of this practice is that toxicity data are nearly always determined for individual chemicals, whereas in real life numerous molecules or classes of molecules coexist in waters with the possibility of multiple interactions.

Among these toxic compounds (including numerous persistent organic pollutants), many are not yet analytically accessible or are analyzable only at exorbitant cost. Thus, it is necessary to develop other strategies to assess the degree to which a given ecosystem is impacted or not by toxic contaminants. In attempting to fulfill this aim, "generalist" biomarkers can reveal the integrated ecotoxicity of complex mixtures, particularly physiological markers linked to the growth and reproduction of organisms.

At the end of the 1990s, several books established the state of the art of biomarker methodology, though the use of biomarkers remains comparatively marginal in ecological risk assessment, still nowdays for many reason, which will be further presented in the book.

Anthropogenic activities are responsible for the environmental input of many classes of chemicals through industrial sources, domestic and urban effluents, and diffuse sources linked to agriculture. The main categories of contaminants include both organic [petroleum hydrocarbons, polychlorobiphenyls (PCBs), pesticides, etc.] and inorganic (metals and nonmetallic elements) compounds. These compounds were studied as soon as ecotoxicology appeared as a specific branch of environmental studies, whereas emerging contaminants have become a topic of concern more recently, even though some of them have been present in the environment for years.

Emerging contaminants include pharmaceutical and care products, alkylphenols, brominated flame retardants, perfluorinated organic compounds, and nanoparticles.

Depending on their physical characteristics, three main categories may be distinguished among chemical wastes: solids, liquids, and gases. Each category corresponds to one of the compartments of our physical environment: lithosphere, hydrosphere, atmosphere. However, it is impossible to describe chemicals entering our environment as continental, aquatic, or atmospheric contaminants since many exchanges occur between these compartments.

Whatever the point of entrance of a given substance into the environment, an important fraction may be carried over what may be a significant distance as a result of water and air circulation. As a consequence, even polar environments are not spared, and in a charismatic species such as the polar bear, increasing levels of persistent organic pollutants are well documented, with possible ecotoxicological effects at the population level [1].

Even if contaminants are distributed on a worldwide scale, dilution in air or water masses increases with distance from the contamination source. This contamination gradient is the primary factor controlling contaminant uptake into organisms (Figure 1). Environmental conditions influence the transformation of many chemicals through chelation, hydrolysis, photodegradation, biodegradation, etc. However, some degradation products of contaminants are not less toxic than the initial molecule, sometimes being even more noxious. Many toxicants are able to cross biological membranes but these membranes and associated structures can act as barriers to contaminant entry (Figure 1). For instance, metal speciation and therefore dissolved metal bioavailability may be modified through ligand secretion into the external medium or by precipitation of dissolved metals as microcrystals of metal sulfides onto the cell surface.

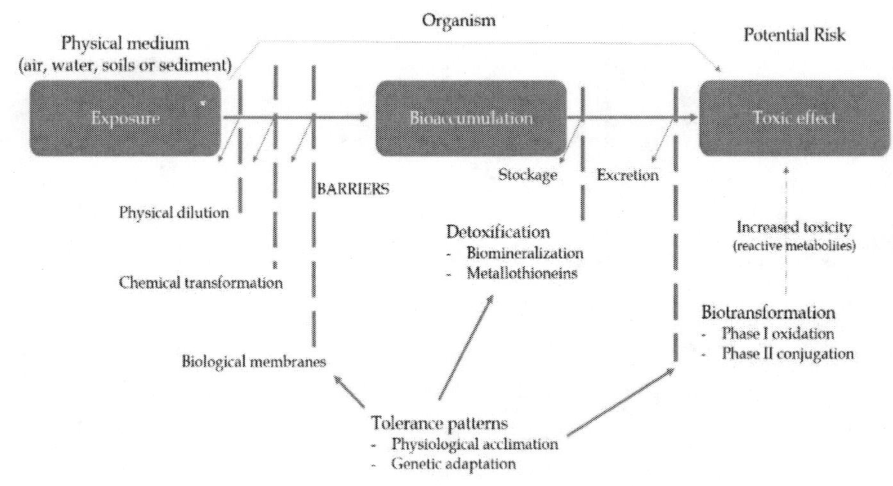

Figure 1. The ecotoxicology triad.

Secretion of exudates by a variety of organisms (bacteria, plants, animals) can involve a great variety of compounds. Subtle changes in the charge and types of reactive groups in such secretions can interfere markedly with their metal binding characteristics and consequently the biological uptake of the metal.

Another mechanism of limiting contaminant uptake is the existence of impervious extracellular barriers such as cuticles, integuments, tests, shells, and scales that contribute to reduce the cell epithelial surface available to contribute to transepithelial transport [2].

Once incorporated into an organism (Figure 1), contaminants can be either stored in tissues or excreted. Storage in intra- or extracellular compartments does not necessarily result in a toxic effect in organisms. For instance, metal detoxification is efficient in numerous organisms. It may be based on the synthesis of metallothioneins (MTs), a family of metalloproteins able to sequester metals via metal binding to their constituent thiol groups, thus blocking any interference between the metals and enzymes that would otherwise result in subsequent enzymatic activity impairments. MT induction is the most common toxic metal defense mechanism in vertebrates. It is also present in most biological taxa [3, 4], but among invertebrates, the major mode of metal detoxification is metal biomineralization in various types of cellular inclusions [5]. It is only when the metal-binding capacity of these ligands is overwhelmed that metal toxicity can occur.

On the contrary, processes responsible for excretion are not systematically free of noxious effects on organisms. Biotransformation of certain organic pollutants [polycyclic aromatic hydrocarbons (PAHs), PCBs] is organized into two phases (Figure 1). Phase I reactions consist of oxidation, reduction, and hydrolysis processes. Phase II enzymes serve to link metabolites from phase I with endogenous substrates, increasing their water solubility and thereby facilitating their excretion. However, phase II biotransformation sometimes leads to reactive metabolites, the interactions of which with cellular macromolecules can engender toxicity [6]. Biotransformation is followed by phase III leading to the elimination of metabolites by the multixenobiotic transport system [6].

The activity of biotransformation enzymes (such as cytochrome P450 enzymes, including ethoxyresorufin O-deethylase involved in phase I; glutathione S-transferase involved in phase II) or MT concentrations are some examples of biomarkers that have been proposed to assess the exposure of organisms to contaminants present in their environment.

In addition to inducing MT synthesis or activating cytochrome P450 enzymes, metals, PCBs, and PAHs can increase oxidative stress by increasing the concentrations of reactive oxygen species naturally present in organisms. Cytotoxicity can occur, including lipid peroxidation and DNA damage, but the degree of such damage depends on the efficiency of enzymatic (superoxide dismutase, catalase, glutathione peroxidase, etc.) and nonenzymatic defenses. If DNA damage induced by metabolites resulting from contaminant biotransformation is not adequately repaired by specialized nuclear enzymes, this can lead to an erroneous expression of the genome, including the activation of oncogenes, which constitutes the first step of the transformation of a normal cell in a tumoral cell [7].

As an indicator of neurotoxicity effects, acetylcholinesterase (AChE) activity has been initially considered a specific biomarker of exposure to organophosphate and carbamate pesticides. More recently, however, other groups of chemicals present in the marine environment including metals, detergents, hydrocarbons, and also cyanobacterium toxins have been shown to affect AChE activity. This lack of biomarker specificity poses a problem for environmental management.

Although biomarkers are able to reveal the presence of contaminants, and subsequent changes in the biology of organisms, any lack of specificity in their response reduces the likelihood of precise targeting of a particular contaminant, thereby affecting management decisions to reduce contamination and its impacts. To date, only a few biomarkers seem really specific: δ-amino levulinic acid dehydratase inhibition in blood able to reveal lead contamination, bile fluorescent compounds for petroleum hydrocarbons [8], and imposex in gastropod mollusks in response to TBT contamination. However, less specific biomarkers are also interesting environmental management tools as general responses to the degradation of environmental conditions, and they are still important in assessing the health status of a given medium exposed to chronic or acute (e.g., oil spill) pollution pressure. Among these biomarkers, stress proteins, which contribute to cellular protection and are highly conserved throughout evolution from bacteria to humans, can provide information on a large spectrum of environmental stress [7].

Histological alterations generally result from the integration of biochemical and physiological changes that may be caused by various chemical contaminants [7].

Furthermore, a variety of nonspecific biomarkers are important because they are involved in growth and development and contribute to the success of reproduction with possible ecological consequences on population sustainability and ecosystem functioning when key species are impacted.

Generally, based on relative considerations to the benefit of specific, less specific, and general biomarkers, it is recommended to date to use biomarkers in a battery for ecological risk assessment, as recommended, for instance by Anderson and Lee [8] and Thain and colleagues [9] in oil spill risk assessment.

Generally, biomarkers have been classified as biomarkers of exposure, effect, and susceptibility [10, 11]. However, the definitions of these classes vary depending on different authors and for these reason certain ecotoxicologists prefer the terminology proposed by De Lafontaine, contrasting biomarkers of defense and biomarkers of damage.

Biomarkers of defense include MTs, phase I, II, and III enzymes evoked above, as well as antioxidant defenses [6] and stress proteins [6]. These defense mechanisms have a positive impact on the health of biota, allowing the survival of organisms in a degraded environment. In highly contaminated zones, many plant and animal species are indeed able to cope with the presence of potentially toxic substances [6].

On the other hand, development of tolerance through physiological acclimation and genetic adaptation can induce energy and fitness costs [6]. Biomarkers of damage reveal more or less severe biological impairments, potentially responsible for detrimental effects on reproduction or even survival.

The importance of toxic effects depending on the degree of environmental contamination is quantified using a dose–effect relationship. The lowest doses do not induce any noxious effect, but with increasing doses biological impairments are progressively enhanced.

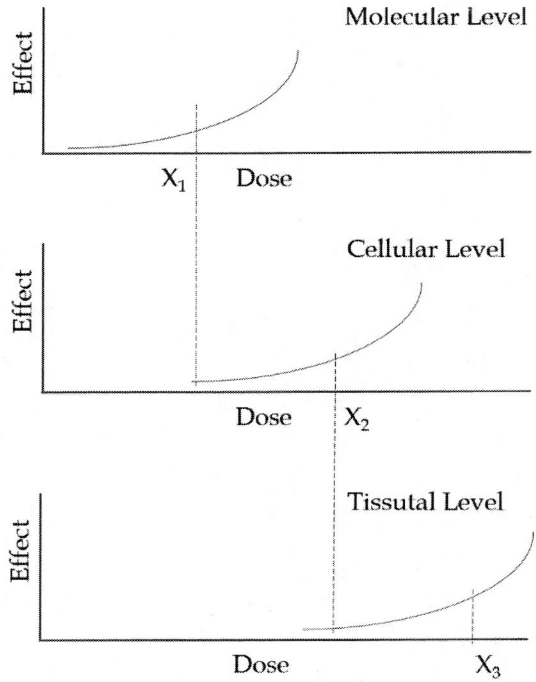

Figure 2. Biomarkers of damage: progression of the dose–effect relationship according to the level of biological organization.

In Figure 2, is shown the theoretical dose–effect relationship for different levels of biological organization. The curve is limited to the domain of low doses to show the first observed effects or initial effects. At the molecular level, the initial effect is observed at a dose X_1 that is lower than the dose X_2 able to induce a cellular effect, this in turn is lower than X_3, and acts at the tissue level. The same argument can be expanded to the level of organs, individuals, populations, etc. This scheme highlights that the lower the level of biological organization, the more sensitive the biological response will be. The rationale for this is quite evident: if only a few molecules have suffered a toxicant effect, cell functioning will not be significantly disturbed; if only a few cells are no longer functional within a whole organ, the function of this organ will still be efficient.

They would appear to have the potential to be particularly useful in a management scheme to prevent any pollution effect, because responses of biomarkers of damage are so sensitive at the lowest levels of biological organization. However, the use of such low level biomarkers brings with it a serious risk of a false positive if they are used as a warning signal for impairments at the level of communities or ecosystems, because organisms have very efficient mechanisms of regulation and repair. This is even true for biomarkers of defense since this type of biological response shows that the organisms are coping actively with environmental degradation. For example, Chapman [12] recommends the use of biological models more representative of the communities or ecosystems under examination than organisms classically used in biomonitoring programs or laboratory tests. It is generally admitted that protecting the most sensitive species within an ecosystem results in the protection of the whole community. This notion of susceptibility is not so simple.

Reproduction and development of juveniles are commonly used as endpoints when assessing interspecific susceptibility to chronic toxicity, because these life traits are considered equally relevant in all species. This hypothesis was tested in two nematode species exposed to copper [13]. Despite juvenile survival, duration of juvenile and reproduction periods, and daily reproduction rate being more affected in one species, fitness (which was defined by these authors as the population growth rate) was identically reduced in both species.

Species most commonly used as biological models in ecotoxicology are representative of the water column, whereas it is well established that sediments and soils are the main stores for a large majority of contaminants entering the environment. The choice of the most relevant species for the determination of biomarkers is crucial and it is based on considering the different objectives of conservation programs: ecosystem functioning, biodiversity integrity, survival of charismatic species, etc.

Responses to pollutants at different levels of biological organization (Figure 3) is represented by considering the latency between exposure and the occurrence of the effect on the X axis, and the degree of ecological relevance on the Y axis, in the case of fish. Molecular effects that are the most sensitive (Figure 3) are also the most precocious. On the other hand, they are mainly toxicological tools for which ecological relevance is poor.

In contrast, population or community responses are obviously relevant to assess the “good ecological status” or “ecological integrity” of water masses [United States’ Clean Water Act (CWA), 1972; European Community Water Framework Directive (WFD), 2000], but effects at these levels become significant only after severe environmental degradation has already occurred, thus leading to expensive remediation processes.

An extreme case provides a striking illustration of the magnitude of remediation problems: the experiences of the Minamata Bay project in Japan [14]. A chemical factory released mercury into this bay from 1932 to 1968, leading to the death of 900 people among more than 2000 affected patients as a result of seafood contamination. The remediation project commenced in 1977 and was completed in 1990 after 1.5 million m^3 of Hg-contaminated sediment had been treated by careful dredging and confined reclamation at a total cost of 48,500 millions yen (equivalent to 650 millions).

Is it possible to reconcile the benefits of biochemical markers and ecological responses? It may be seen in Figure 3 that processes involved in reproduction include a set of responses from the molecular level leading to consequences of reproductive success on the sustainability of populations in ecosystems impacted by anthropogenic activities.

Although it is excessive to consider that the pursuit of toxicological endpoints other than those concerned with reproduction is likely to be a wasted effort [15], it is evident that reproductive success is key for environmental conservation.

The impairments at infra-individual and individual levels that can most probably affect the success of reproduction are depicted in Figure 4. These include endocrine disruption, behavioral changes, energy disturbances, and genetic responses, either adaptive or detrimental.

The problem of endocrine disruption was first realized because of the disastrous ecotoxicological effect of tributyltin (TBT), a compound used in antifouling paints. TBT-mediated imposex has been observed in more than 195 species of prosobranch gastropods worldwide [16]. Subsequent population depletion of such gastropods has been observed in harbors and marinas where many individual snails were presenting morphological symptoms of imposex.

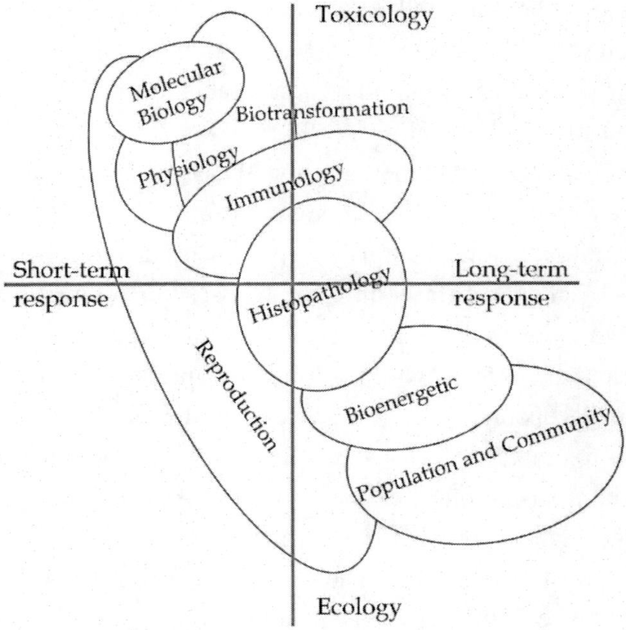

Figure 3. Latency between exposure of fish to pollutants and the occurrence of effects at different levels of biological organization.

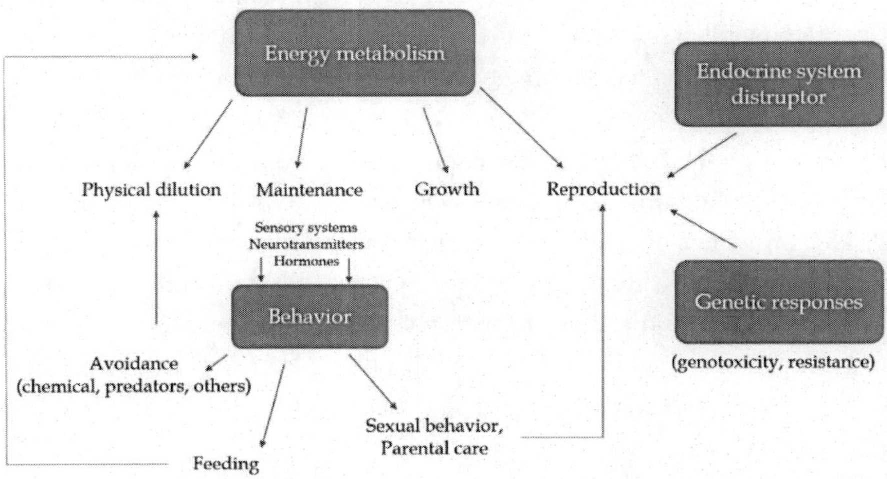

Figure 4. Linkage between effects of contaminants from molecular to population levels via the success of reproduction.

In the case of the dogwhelk (*Nucella lapillus*), population-level effects on other species (barnacles, fucoid seaweeds, hermit crabs) belonging to the same ecological community would be attributable to such a population drop in the affected gastropods [17].

Endocrine glands and the hormones they secrete are not only indispensable to the success of reproduction, but are also involved in the development of organisms, their growth, and their behavior. However, most scientific research, particularly in fish, focuses on interactions between pollutants and male and female sexual hormones. A peculiar topic of concern is that the effects of endocrine disruptors on reproduction are typically subtle, occurring at low doses, in the absence of any other appearance of toxicity.

The spatial distribution of endocrine-disrupting chemicals, particularly steroid estrogens and nonylphenols, is related to the discharge of domestic and industrial wastewaters everywhere in the world [18, 19, 20, 21]. The presence of intersex (male gonads invaded with oocytes) individuals is increasingly documented in bivalves and fish. Natural or xenoestrogens could be a contributory factor in the induction of intersex [22, 23].

A wide variety of anthropogenic, waterborne contaminants can also affect the hypothalamic–pituitary–thyroid axis and its role in development and reproduction, as recently reviewed in teleost fish and amphibians [24, 25]. Impairment of thyroid functioning can influence behavior as neurotoxic effects such as the inhibition of neurotransmitters (AChE, serotonin) have also been observed (Figure 4). Many aspects of behavior can be affected [26, 27, 28]: avoidance of predators or contaminated sediment or other habitat, contributing to the defense and survival of organisms; location of sexual partners and care of juveniles indispensable to reproductive success; feeding behavior and prey capture important for acquiring energy. Thus, behavioral ecotoxicology is potentially useful to link biochemical impairments to population effects.

The success of reproduction is clearly linked to the relative energy allocation of an organism to defense against exposure to chemical stressors, basal metabolism, growth, and reproduction. Organisms obtain their energy from ingested food. For predators, the impairment of foraging activity can lead to a shift toward easily accessible food such as detritus, the energy value of which may be lower. Chemical contaminants can also influence food assimilation through the impairment of digestive enzyme activity. Lastly, prey species can be susceptible to environmental contamination, thus leading to decreased food availability for predators.

Energy analysis can reveal a disequilibrium in energy balance associated with toxic or more general stress. Different energy parameters can be used as biomarkers of pollutant effects. These parameters can be linked to macroscopic criteria representative of maintenance and growth (condition indices, size, or biomass increase, etc.) or reproduction (gonadosomatic index, egg production, offspring number, etc.).

For ecological risk assessment, it is necessary to determine to what extent populations may be affected when such adverse effects are revealed (loss of their ecosystem function or even local extinction). Models that can allow extrapolation from individual- and suborganismal-level responses to the population level have been reviewed [29]. Among those, dynamic energy budget models combined with demographic models have been well developed [30].

Exposure to chemicals can lead to DNA damage (Figure 5), the consequences of which may be limited by DNA repair [31]. Mutations frequently have toxic effects, including carcinogenesis, and when affecting germinal tissues, they are inheritable and can also affect future generations, provided that the offspring are viable and able to survive and reproduce. In fact, impairments of germinal cells often result in embryo lethality or early death of the progeny. From an ecological point of view, it is questionable if these precocious deaths can impact the fate of populations [10, 7]. In some cases, mutations can confer a selective advantage leading to the selection of resistant genotypes.

Chronic exposure to chemicals can exert a selection pressure leading to the presence of resistant genotypes in organisms living in impacted areas. The acquisition of tolerance is particularly well documented for pesticide-exposed insects [32], but other classes of contaminants (metals, PAHs, PCBs) can be responsible for the predominance of resistant genetic patterns in bacteria [33], plants [6], invertebrates [34], and vertebrates [35, 6]. In contaminated areas, an increased frequency of resistant genotypes has often been reported, allowing the maintenance of DNA integrity associated with the duplication of specific genes (Figure 5).

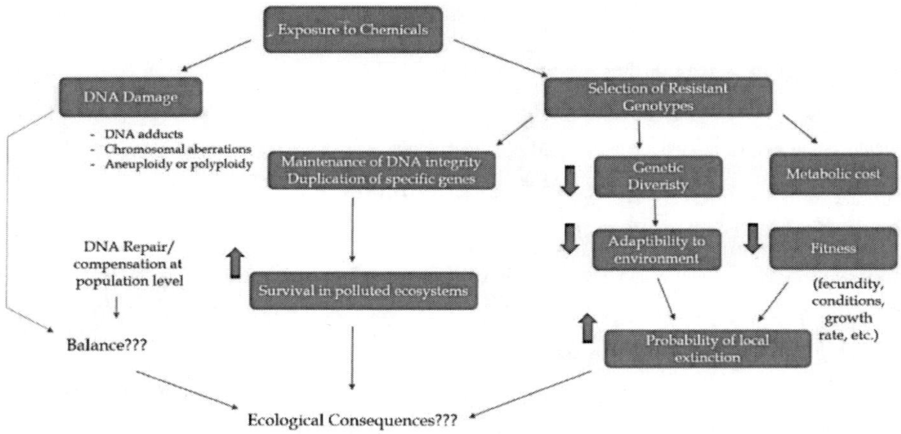

Figure 5. Genetic responses to chemical exposure: DNA damage versus selection of resistant genotypes.

However, negative consequences of being resistant may be observed, such as decreased fitness and decreased adaptability to new environments or stressors, thus increasing the probability of local extinction.

Biomarkers are available as crucial tools in ecotoxicology, because they can be used as early warning signals of environmental change before the onset of irreversible damage at the population level. Syntheses published at the turn of the century [36, 37, 38] suggested that scientists were then ready to transfer the methodology of biomarkers to end users in charge of environmental biomonitoring. A decade later, certain biomarkers are used to assess the health status of aquatic environments in different parts of the world. However, this use is generally limited to a relatively small number of more or less specific biomarkers, the worst counterexample being the WFD—a very important regulation aiming at the protection of aquatic environments from the river source to the seashore—which totally ignores the use of biomarkers despite the efforts of European scientists to demonstrate the relevance of biomarkers as tools for the implementation of the WFD [39, 40, 41].

Independently of regulatory frameworks, many important studies have demonstrated "the usefulness of applying a large array of various combined biomarkers at different levels of biological organization, in assessing the toxic effects of a mixture of pollutants in a natural aquatic environment" (Huadi River, a tributary of the Pearl River, China) [42].

In the Bay of Cadiz, biomarkers determined in caged clams *Ruditapes philippinarum* allowed assessment of chemical exposure and sediment quality [43].

In the Río Champotón (southwestern Mexico), a set of biomarkers determined in a native fish *Astyanax aeneus* was shown to be a sensitive and effective tool for identifying periods of environmental conditions adverse to fish health [44].

Several problems contributing to limit the use of biomarkers have been recognized: the problem of confounding factors, e.g., [9, 45], the question of a reference site, and the lack of ecological relevance [46].

The problem of confounding factors was well conceptualized by Cairns [47]. When a biological parameter is highly fluctuating, the occurrence of a stress may be concealed by natural fluctuations. On the other hand, when background values are relatively stable, any change due to contamination factors is easily revealed. As already mentioned by Kalman and colleagues [48], "The question of confounding factors is well mastered in biomonitoring programs based on the determination of contaminants in the tissues of bioaccumulators such as the bivalves used in the 'Mussel Watch'–type programs."

For many aspects of environmental monitoring, our present state of knowledge and the insufficiency of background data available mean that the use of a reference site for comparison is essential. However, to date, with the worldwide dispersion of contaminants evoked above, pristine areas have disappeared and, at best, reference sites can be chosen in only a few places that remain comparatively clean. To choose a reference site, geographical proximity and similarity in terms of temperature, granulometry, and organic content of sediment, salinity regime (in estuaries), etc., are mandatory to mitigate the importance of confounding factors. This is not an easy task, as described, for instance, in estuaries [49].

Potential reference estuaries with low perceived anthropogenic pressure are generally small, whereas the human activities responsible for the presence of many chemicals in the environment have historically developed on the banks of larger main watercourses. This does provide a potential problem when trying to eliminate comparative differences resulting from hydrodynamic differences between the estuaries under comparison. Even in the less fluctuating conditions of a freshwater biomonitoring program, the interpretation of fish biomarker results is strongly influenced by the selected reference system [50].

The addition of more than one reference site into any comparative study, however superficially attractive, has significant resource implications. Associated with the need for temporal surveys instead of spot sampling techniques and the development of the need to analyze a battery of biomarkers, methodology involving biomarkers is not always as initially claimed: sensitive, simple, and cost-effective.

Even despite this complexification, the biomarker methodology to be proposed to end users—although efficient in assessing chemical exposure, sediment quality, and the toxic effects of mixed pollutants— still fails at predicting chemical risk at supra-individual levels [46].

The development of an integrated indicator framework using biological effect techniques remains key to improve the risk assessment of contaminants in aquatic ecosystems [9]. Ecologically relevant biomarkers such as lysosomal integrity, immunotoxicity, endocrine disruption, behavior, energy metabolism, and genomic biomarkers appear to be promising candidates to fill the gap existing between sub-organismal and organismal responses to stress and effects occurring at higher levels of biological organization.

The aim of the present book is to review biomarker research that examines the effects of contaminants using an integrative approach. New approaches in functional genomics and bioinformatics can help discriminate individual chemicals, or groups of chemicals among complex mixtures that may contribute to adverse biological effects. By using these recent methodologies, it could be useful to put light on the molecular evolution of the biomarkers, which role and functionality can be better understood based on exploring the relative evolution pathways in several aquatic organisms.

Chapter 2

CLASSES OF BIOMARKERS

2.1. DEFINITIONS

In the past nearly 35 years, several definitions of biological markers have been published. The historical development of the biomarker approach has been closely related to advances in medicine and biology of vertebrates. Biological markers were classified as exposure, effect, and susceptibility biomarkers. Moreover, in the publications of the National Research Council [51, 52], the authors highlighted that biological markers can be simultaneously used for biological monitoring and for monitoring of health.

According to McCarthy and Shugart [53], "biological markers are measurements at the molecular, biochemical, or cellular level in either wild populations from contaminated habitats or in organisms experimentally exposed to pollutants that indicate that the organism has been exposed to toxic chemicals, and the magnitude of the organism's response to the contaminant."

The definition was generalized by Depledge [54]: a biomarker is "a biochemical, cellular, physiological or behavioral change which can be measured in body tissues or fluids or at the level of the whole organism that reveals the exposure at/or the effects of one or more chemical pollutants."

In September 1994, the journal Ecotoxicology presented four reviews on the role of the biomarkers in environmental assessment, as carried out with invertebrates [55], vertebrates [56], terrestrial plants [57], and populations and communities of invertebrates [58]. These articles were required by the European Foundation for Science (ESF) to understand to what extent biomarkers could be used to evaluate environmental damage and to formulate possible rules to control any such damage.

Later, Van Gestel and Van Brummelen [59] attempted a redefinition of the terms biomarkers, bioindicators, and ecological indicators, by calling on previous work published in Ecotoxicology in 1994 when Lagadic and colleagues [58] made a clear distinction between biomarkers and bioindicators and restricted the use of the term "biomarker" to the sublethal biochemical changes resulting from individual exposure to xenobiotics. However, this reductionist definition was not generally accepted [60, 39], with many scientists voicing their concern about not neglecting responses (e.g., physiological, behavioral) that could be used in risk assessments involving a change in scale of biological organization from the individual to the population.

According to Van Gestel and Van Brummelen [59], a biomarker is defined as any biological response to an environmental chemical contaminant at the infra-individual level, measured in an organism or its products (urine, feces, hair, feathers, etc.), indicating a change compared to the normal state and which cannot be detected in a healthy organism.

The term bioindicator should be restricted to an organism providing information on the environmental conditions of its habitat by its presence or its absence or its behavior.

The concept of specific biomarkers (responding to metal pollutants, or to organics or to any defined pollutant) led to the definition of damage and defense biomarkers (Table 1) put forward by De Lafontaine and colleagues [61].

From the 1970s, great developments in biochemistry and molecular toxicology made it possible to progress quickly in our knowledge of the mechanisms of the toxicity of xenobiotics, mainly with mammalian models. Thereafter, significant specific biochemical effects were highlighted in species exposed to some contaminants, particularly in birds, fish, and mollusks considered as being of ecological interest.

Table 1. Types of biomarkers according to De Lafontaine and colleagues [61]

Defense Biomarkers	Damage Biomarkers
Ethoxyresorufin O-Deethylase	AChE Activity
Fluorescent Aromatic Compounds in	Vitellogenin
Fish Bile	Lysosomal Membrane Stability
Phase II Enzymes	Thiobarbituric Acid Reactive Substances
Phase III Enzymes	DNA Damage
Metallothioneins	
Enzymatic and Nonenzymatic	
Antioxidant Defenses	
Heat Shock Proteins	

Nowadays, the majority of the examples concern the aquatic environment, particularly the marine environment, which is the final receptacle of chemical pollutants.

Well-known biomarkers, which have been recognized in laboratory and environmental studies, have been called "core biomarkers" [62]. Such core biomarkers include the stability of the lysosomal membrane (measured by the neutral red retention time), inhibition of acetylcholinesterase (AChE) activity, metallothionein (MT) concentration, ethoxyresorufin O-deethylase (EROD), and the fluorescent metabolites of the bile [fluorescent aromatic compound (FACs)].

Based on the last review performed by Kroon and colleagues [11], like previously mentioned, the definitions of biomarkers are rather diffuse [63, 64, 65]. For example, many definitions of biomarkers of exposure include both body burden as well as early response to contaminants. Similarly, definitions for biomarkers of effects generally refer to biochemical and physiological alterations resulting from exposure to a contaminant, but sometimes also include changes ranging from molecular, behavioural, and up to ecosystem level.

In this book is presented the alternative system of defining the different biomarkers, which was developed by Kroon and colleagues [11] and represented by the following clear working definitions of biomarkers for their study, in order to ensure consistency with the global literature, like it was on fish health assessment:

- Bioaccumulation markers: Analytical/chemical indicators inside an organism or its products (also referred to as body burden) [63],
- Biomarkers of exposure: Markers which indicate an early biochemical response has occurred following exposure of an individual or organism to a contaminant [64],
- Biomarker of effect: Measurable biochemical, physiological or other alterations within tissues or body fluids of an organism that can be recognized as associated with an established or possible health impairment or disease [64, 63],
- Biomarker of susceptibility: Inherent or acquired ability of an organism to respond to the challenge of exposure to a specific xenobiotic substance, including genetic factors and changes in receptors which alter the susceptibility of an organism to that exposure [63].

Anyway, it is important that most of the authors are specifically focused on bioaccumulation markers, biomarkers of exposure, and biomarkers of effect.

2.2. Bioaccumulation Markers

The most common analytical techniques to measure metal bioaccumulation in fish tissues [11] were inductively coupled plasma mass spectrometry (ICP-MS), inductive coupled plasma atomic emission spectrometry (ICP-AES) (also referred to as inductively coupled plasma optical emission spectrometry, ICP-OES), and atomic absorption spectroscopy (AAS).

Bioaccumulation factors (BAFs) for individual fish species, calculated as the metal concentration in the organism (or tissue) divided by that in the sediment or water, were reported in only six out of the 118 studies [11]. Bioaccumulation factors for sediment (BAFsed) and water (BAFwater) were mostly determined for Cd, Cu and Zn, with BAFsed values for Cd ranging from 0 in *Cathorops spixii* [65] to 521 in *Liza microlepis* [66], for Cu from 0.13 for *Stromateoides argenteus* [67] to 304 for Liza saliens [68], and for Zn from 0.21 in *Johnius belengeri* [69] to 447 in L. saliens [68].

The values for BAFwater for Cd ranged from 1,048 in *Cynoglossus sinicus* to 21,333 in *Trypauchen vagina*, for Cu from 987 for *Argyrosomus argentatus* to 2,242 in *Setipinna taty*, and for Zn from 371 in *Johnius belengeri* to 1,651 in *Leiognathus rivulatus* [69].

The high values for BAFsed and BAFwater for Cd, Cu and Zn suggested a high bioaccumulation potential for these metals, in fish muscle and liver tissue in particular. Of the seven contaminants of concern, bioaccumulation of Cd, Cu, Pb and Zn in fish tissues was most often examined, in both field and laboratory studies. In contrast, bioaccumulation of Al and Se was examined much less frequently, or in the case of Ga not at all. Metal bioaccumulation, however, does not always reflect environmental metal concentrations [70, 71, 72, 73], and can vary with fish species [71, 74, 75, 76], tissue [70, 71, 73, 77], and life history stage [70, 75], with exposure pathways [75, 78, 79], with season [80, 81], and can be influenced by metal speciation and bioavailability [73, 76, 82]. Hence, when using biomarkers to assess fish health, metal bioaccumulation should be used in conjunction with other biomarkers to indicate whether biological responses are related to bioaccumulation

2.3. Biomarkers of Exposure

Generally, the biomarkers of exposure to single or multiple contaminants with similar modes of action show an early response to contaminants and are typically specific to a particular class of contaminants, e.g., biliary fluorescent aromatic compounds (FACs) for exposure to oil, or the induction of the egg-yolk precursor protein vitellogenin (VTG) for environmental estrogens [83].

One of the most common biomarkers in this category is the measurement of the induction of a cytochrome P450 1A (i.e., CYP1A) at the messenger RNA, protein, or catalytic activity levels, as biomarkers of exposure to components of oil [84]. Furthermore, the presence of some PAH metabolites excreted in the bile of fish (e.g., bile FACs) has also been used successfully to track the temporal and spatial effects of oil pollution [85]. Oil biomarkers are particularly important as some components of oil can be rapidly metabolized in aquatic organisms or undergo environmental breakdown. However, elevations in oil exposure indices such as CYP1A activity have not been linked to higher level biological effects [84]. Accordingly, measures of CYP1A as well as biliary FACs would have most predictive power in a regulatory context when used as part of a suite of biomarkers, including those discussed below that are more intricately linked to physiological and higher-level effects.

Induction of vitellogenin (VTG) in male or juvenile fish is a commonly used biomarker of exposure to environmental estrogens, e.g., [86, 87, 88]. This induction can be measured both at the protein [87, 88, 89] and at the mRNA level. Increased VTG levels have been linked to pharmaceuticals in sewage [87, 88], to overall anthropogenic impact [90], as well as to phytoestrogens in pulp and paper mill effluents [86].

Metallothionein (MT) has known roles in routine physiological processes involving essential metals and in detoxifying non-essential metals such as cadmium [91]. Different MT isoforms can also be responsive to oxidative stress [92]. MT mRNA levels in fish have been shown to vary along a contamination gradient [93], and also have been shown to correlate with bioavailable metals in fish caged along contamination gradients [94]. MT mRNA levels were a highly sensitive indicator of laboratory cadmium exposures in Coho salmon [95]. Interestingly, another study of Coho salmon indicated that measuring MT mRNA levels in the olfactory system of salmon was more sensitive than in the liver, suggesting tissue-specific differences in MT induction [96].

Acetylcholinesterase (AChE) inhibition has long been used as a biomarker for exposure to, as well as effects of, carbamate and organophophorus pesticides in both fish and invertebrates [97], although it is not always as effective a biomarker for molluscs due to avoidance behaviours [98]. AChE inhibition is a rare and in some ways ideal biomarker as it indicates both exposure and effects. Inhibition can be linked to both decreased swimming stamina and survival, although the exact relationships between these linkages tend to be species specific. Exposure to pesticides and AChE inhibition are also thought to be contributing to the poor recovery of some Pacific salmon populations [99].

Below are presented the all the biomarkers of exposure and their relative characteristics [11] in terms of the specific role and involvement in several molecular and physiological processes.

2.3.1. Biotransformation Enzymes: Phase I

Biotransformation is the process by which an organism alters a xenobiotic compound to enable its excretion. Phase I reactions are the major pathway for the biotransformation of lipophilic compounds and include oxidative, reductive, and hydrolytic reactions where a polar group is either introduced or unmasked (*via* the addition of an oxygen atom), rendering the xenobiotic molecule less active and more water-soluble and allowing it to be more readily excreted.

Most Phase I reactions are catalysed within the microsomal mono-oxygenase (MO) enzyme system and are dependent on the heme protein cytochrome P450 (cyt P450). These heme proteins are located predominantly in the liver, but also present in other fish organelles and tissues [64].

Cyt P450 isozyme induction is triggered when specific xenobiotic compounds bind to the aromatic hydrocarbon (Ah) receptor protein complex and the heat-shock protein 90 (HSP 90), causing HSP 90 to be released.

Exposure to certain xenobiotic compounds, such as PAHs and PCBs, can show a strong and highly specific induction in the cyt P450 isozymes and a significant corresponding elevation in total cyt P450 levels.

The use of total cyt P450 as a biomarker, however, has been somewhat superseded by techniques that target specific isozymes such as cytochrome P450 1A (CYP1A) which allow more specific discrimination of causal agents.

Cytochrome P450 1A is a class of cyt P450 isozymes responsible for the biotransformation of a number of compounds including PAHs, halogenated aromatic hydrocarbons (HAHs), PCBs, and dioxins. CYP1A protein levels respond to specific xenobiotic compounds and are generally more responsive to contaminants than other CYP isozymes. CYP1A protein levels can be determined cheaply via immunological tests such as enzyme-linked immunosorbent assays (ELISA) or via histochemical techniques. Xenobiotic related elevations in CYP1A proteins are usually preceded by an increase in CYP1A mRNA, which may also be used as a biomarker.

Four different methods have been used to assess cytochrome P450 1A (CYP1A) activity as a biomarker of fish health when exposed to metal contamination; all of these studies have used sediment or water as the toxicant pathway. CYP1A activity has been measured in liver, gonads, gills and skin. CYP1A activity in gonads and skin did not change in response to metal contamination [100].

Ethoxyresorufin O-deethylase (EROD) activity is a measure of catalytic response in fish, and is a sensitive indicator of the inductive response of the cyt P450 system. EROD activity is indicated by changes in the fluorescence of the metabolic product resorufin. Aryl hydrocarbon hydrolase (AHH) catalytic activity can be used as a measure of the response of the CYP1A isoenzyme, by determining the hydroxylation of benzo[a]pyrene.

Both EROD and AHH are sensitive biomarkers when measured in the livers of fish, especially when used in conjunction with other biomarkers such as CYP1A and CYP1A mRNA. The other sub units of the cytochrome 450 system are measured using the following assays: pentoxyresorufin-O-depentylase (PROD) measures the catalysis of cytochrome P450 CYP2B, 7-benzyloxyresorufin-O-depentylase (BROD) measures the catalytic activity of the cytochrome P450 subfamily CYP3A and 7-methoxyresorufin-O-methoxyresorufin (MROD) measures the catalytic activity of the cytochrome P450 CYP1A2.

2.3.2. Biotransformation Enzymes: Phase II Enzymes and Co-Factors

Phase II biotransformation involves the attachment (conjugation) of a polar molecule to the xenobiotic parent molecule. The resultant molecule is more water soluble, of higher molecular weight and is more effectively excreted. Depending on the type of xenobiotic compound, certain molecules may be directly metabolised by the phase II system; others may require biotransformation *via* phase I enzymes first.

Reduced glutathione (GSH) and oxidized glutathione (GSSG) provide the major pathways for conjugation of electrophilic compounds and metabolites. GSH plays a key role in the detoxification of xenobiotic compounds by reacting with compounds, replacing hydrogen, chlorine and nitro groups among others. Thiol status is the ratio between reduced and oxidised glutathione (GSH/GSSG) and has been used as an indicator of oxidative stress.

Glutathione S-transferase (GST) is a catalyst required for the conjugation of electrophilic compounds by GSH [101]. GST provides essential functions in intracellular transport as well as defence against oxidative damage and peroxidative DNA products.

2.3.3. Oxidative Stress Parameters

Oxidative stress parameters are a suite of measures of the oxygen toxicity being suffered by an organism. Oxygen toxicity is caused by reactive oxygen species (ROS) and oxygen free radicals, and is a result of either increased production of these oxidising species, or a significant decrease in production of antioxidants (such as glutathione).

Severe oxygen toxicity can lead to enzyme deactivation, lipid peroxidation, DNA damage and eventually apoptosis or cell necrosis.

ROS may be produced endogenously, or as a result of the reduction of xenobiotic compounds such as aromatic diols, quinones and hydroxyl-amines, nitroaromatics, bipyridyls and some transition metal chelates [64].

Superoxide dismutase (SOD), catalase (CAT), and glutathione peroxidase (GPOX), are groups of antioxidant enzymes that form part of a cellular defence system that inhibit and detoxify oxyradical formation. SODs are metalloenzymes that catalyse the reaction of superoxide anions into hydrogen peroxide (which is also an ROS). Resultant hydrogen peroxide is then detoxified by CAT and GPOX. CATs enable the conversion of hydrogen peroxide to release molecular oxygen and water and are involved in fatty acid metabolism as well as detoxifying xenobiotic ROS.

GPOXs facilitate the reduction of a range of peroxides (unlike CAT which only acts on hydrogen peroxide) to alcohols. Where CAT uses a donor hydrogen peroxide molecule in the reduction of another, GPOX requires an additional reductant to achieve this (such a selenium-dependant tetrameric cytosolic enzyme), and commonly utilises reduced glutathione (GSH) as a cofactor. GPOX is crucial in protecting cell membranes from damage via lipid peroxidase (LPOX).

Glutathione reductase (GRED) maintains homeostasis between the reduced and oxidised forms of glutathione, namely GSH and GSSG, particularly under oxidative stress conditions. GRED uses the oxidation of NADPH to NADP+ to catalyse the reduction of GSSH to GSH. NADPH levels can be measured spectrometrically and can therefore be used as a proxy for GRED activity.

It is important to notice that non-enzymatic antioxidants such as vitamins have been used as biomarkers although relevant studies are very scarce [64]. Lipid peroxidation (LPOX) is the oxidation of polyunsaturated fatty acids as a consequence of oxidative stress, and can be used as biomarker by quantifing and measuring the degradation products such as aldehydes, acetone and malondialdehyde.

2.3.4. Stress Proteins, Metallothioneins and Multixenobiotic Resistance

Stress proteins protect and regenerate cells in response to stress and harmful conditions. The group consists of heat shock proteins (HSPs), which respond (by increasing synthesis) to heat and other physical and chemical stresses, the glucose-regulated proteins (GRPs) which respond to oxygen or glucose deficiency, and the stressor-specific stress proteins, which include metallothioneins (MTs) and heme oxygenase proteins.

Metallothioneins (MTs) are a family of proteins essential for the regulation, sequestration and detoxification of metals including Cu, Zn, Cd, Hg and have been shown to be induced by Co, Ni, Bi and Ag. MTs act by intercepting and binding metal ions as they are being taken up by the cell and also by the removal of metals by non-thionein ligands. MTs can be measured by quantitative assays of the protein via IEC-AAS, metal substitution assays and polarographic and immunochemical techniques. In fish, MT activity is greatest in tissues that are most closely involved in uptake, storage and excretion (such as the gills, liver and small intestine). MTs are one of the most studied biomarkers for assessing the effect of metal contamination in fish. Multixenobiotic resistance (MXR) prevents the accumulation of both xenobiotic and endogenous compounds inside the cell, by removing them via an energy-dependent transport protein, Transmembrane P-glycoprotein (PGP).

2.3.5. Haematological Parameters

Haematological parameters offer the ability to conduct nonlethal or non-destructive tests. Haematological parameters may be less chemical specific than some other biomarkers, but provide other insights into the general health and physiology of the organism. A number of haematological parameters have been used to monitor pollution induced stress in fish, including plasma enzyme activity, erythrocyte counts, haemoglobin content and haematocrit.

2.3.6. Immunological Parameters

Immunological parameters have been used often as biomarkers in fish exposed to metal contamination, with the most common ones being proliferating cell nuclear antigen (PCNA) [102, 103, 104], calcium binding proteins (CBP), cell respiratory burst (RB) and total complementary activity of serum (TCAS) [105, 103]. Additional immunological parameters that have been used include % leukocytes and % thrombocytes [106], stability of lysosomes 1 and 2 [107, 108], lysozyme activity [109, 110], macrophage aggregate activity (MAM), macrophage aggregates (MA), proteomics serum (Pro-S), cytokine transforming growth factor (β1 9TCF-b1), melanomacrophage centres (MMC), thymus volume (TV) and cell viability (CV).

2.3.7. Neurotoxic Parameters

Cholinesterase (ChE) are enzymes relevant to neural functions; those with a high affinity for acetylcholin (AChE), and those with affinity for butyrylcholin (BChE) [64]. Fish muscle contains both, while fish brains only contain the former. Nine studies used ChE as a biomarker in fish exposed to metals under field conditions [107, 111, 112, 113, 114, 115, 108], with variable reported responses. Most studies measuring ChE activity in the liver did not identify a response in fish exposed to contaminated sediment [107, 110, 108], while other studies showed a mixed or negative response [112]. The neurotransmitters choline acetyltransferase (ChAT) and serotonin, as well as the enzymes tyrosine hydroxylase (TH), involved dopamine synthesis and the serotonin receptor (5-HT3) have been used as fish biomarkers [116]

2.3.8. Reproductive and Endocrine Parameters

Pollution stress has well-known effects on reproductive and endocrine parameters in fish [117], however, only a few studies specifically examined these biomarkers [11]. Vitellogenin (VTG) is a precursor egg yolk protein found in the blood or haemolymph in females of nearly all oviparous species [118]. VTG levels are highly sensitive to endocrine disrupting chemicals (EDCs), and are affected by exposure to some xenobiotics including PCBs and Cd [64].

2.3.9. Genotoxic Parameters

Genotoxic parameters refer to any changes to the genetic material of an organism caused by exposure to a contaminant. A number of genotoxic parameters have been used to monitor pollution induced stress in fish, including DNA adducts, secondary DNA modifications, and irreversible genotoxic events.

A DNA adduct is a form of DNA damage caused by the binding of a chemical to a segment of DNA. The two studies showed different responses in DNA adducts in *P. flesus* exposed to metals, namely an increase [110] and no change [119]. These inconsistent results in a low number of studies preclude recommendation of this biomarker for fish health assessments in Gladstone Harbour.

Secondary DNA modifications. Secondary DNA modifications including changes in DNA strand breaks, DNA unwinding, as well as programmed cell death (i.e., apoptosis). DNA strand breaks can be assessed using the comet assay and has been examined in 14 studies [107, 120].

2.3.10. Other Biomarkers of Exposure

A range of other biomarkers of exposure have been used to assess the response in fish following metal exposure, including osmoregulatory and respiratory markers, and carbohydrate and aerobic metabolism, but we are not going to elucidate more details about these biomarkers, due to the limited number of studies, which show that yet they have to be confirmed as biomarkers.

2.4. Biomarkers of Effect

Examples of effects biomarkers, i.e., indicators of physiological or biochemical changes as a consequence of exposure, are listed in Table 2. These indices may be direct measures (e.g., DNA damage, AChE inhibition), or indirect measures such as the impact on sub-cellular lysosomes. Theoretically, an organism with a quantitative change in an effects biomarker will undergo some loss of fitness. However, these associations can be difficult to demonstrate quantitatively.

Despite their advantages, a major disadvantage of effects biomarkers is that very few can be linked directly to exposures to specific classes of chemicals. However, there are some notable exceptions, including AChE activity, as well as the presence of PAH-DNA adducts (Table 2). However, effects biomarkers can be effectively incorporated with other diagnostic markers of fish health and also with analytical chemistry approaches to provide evidence for the contributions of chemical exposures.

Biomarkers of oxidative stress represent a unique sub-class of effects biomarkers that have been applied inconsistently in aquatic studies. Oxidative stress is part of the ageing process and occurs in all living organisms when reactive oxygen species (or their by-products) cause cell and tissue injury [131, 132]. It is well accepted that exposure to a broad range of environmental chemicals, including pesticides, metals, and PAHs increase the level of cellular oxidative stress in aquatic organisms [131, 132, 133, 134, 135, 136, 137]. Aquatic organisms contain a full complement of antioxidant enzymes, which comprise antioxidant defenses, including enzymatic components (e.g., superoxide dismutase (SOD), catalase and glutathione peroxidases) as well as small molecule antioxidants (e.g., glutathione). Since antioxidant defences can be quantitatively altered on exposure to contaminants, this has led to their exploitation as biomarkers in the field [133, 134, 135, 136]. A breakdown in the antioxidant detoxification procedures with the potential for higher order effects has also been observed for metals in bivalves [137].

Elevated antioxidant enzyme parameters associated with increased lipid peroxidation have been observed in the fish, *Fundulus* sp. exposed to complex mixtures of organic chemicals [138]. Similar results have been reported for bivalves [135]. Oxidative stress-associated parameters were particularly sensitive biomarkers of contaminants in receiving waters of a hydroelectric power plant [139].

It is also important to examine several endpoints related to oxidative stress in different tissues, because the response of the organism may be both xenobiotic- and also tissue-specific. For example, [140] compared brown bullhead (*Ameriurus nebulosus*) collected from a system contaminated with PCBs to bullhead collected from a relatively non-contaminated site. Fish from the contaminated site had extremely high PCB concentrations in muscle compared to fish from the non-contaminated site, and SOD activity was increased in the kidney of fish from the contaminated site. However, the activity of another antioxidant enzyme, catalase, was lower in the kidney, and there were differences in the trends of expression of other enzymes. Studies such as these highlight the importance of tissue selection when evaluating oxidative stress in feral organisms.

Protective antioxidant enzymes and their non-enzymatic cofactors (e.g., GSH) can be overwhelmed during high levels of tissue oxidative stress, which can lead to the accumulation of metabolic by-products reflecting oxidative damage. These metabolic by-products include oxidized pigments such as lipofuscin, peroxidized cell membranes and their ensuing membrane breakdown products (i.e., lipid peroxides, malondialdehyde), as well as oxidative DNA and protein damage. These secondary products of oxidative damage can initially compromise the ability of the organism to maintain normal metabolic processes, and under chronic oxidative stress, to the disease states mentioned above. Lipid peroxidation, in particular, has been routinely measured in field studies to reflect chemical-induced oxidative damage.

It is important to note that there are quantitative differences in the normal levels of these antioxidants, as well as their induction capabilities among the various aquatic species [132].

Accordingly, oxidative stress biomarkers need to be carefully validated under controlled laboratory conditions.

Table 2. Examples of biomarkers that have been used to assess aquatic contaminants

Category	Measurement and Indication	References
Biomarkers of biological effects		
Heat-shock proteins (i.e., HSP 90)	Proteins that are induced in tissues of aquatic organisms as a generalized response to stress, including exposure to chemicals, hypoxia, and temperature	[121]
Markers of oxidative stress (heme oxygenase, superoxide dismutase, glutathione, catalase, lipid peroxidation)	Enzymes, cofactors in metabolic products that can be quantitatively altered on exposure to pollutants. Oxidative stress is an adverse cellular and tissue level response to a variety of contaminants and non-chemical stressors	[122, 123, 124, 125]
Markers of lysosomal membrane stability	Sub-cellular organelles containing hydrolytic enzymes that are sensitive to toxicity leading to membrane rupture and leakage (metals)	[126, 127]
Condition indices (hepatosomatic index, gonadal indices)	Decreases in organ weight relative to whole body weight can reflect organ toxicity or disease	[90, 128]
Circulating hormone levels	Levels of circulating hormones can be used to measure normal sexual and reproductive behaviour	[128]
DNA damage measures (induction of DNA repair enzymes, presence of PAH-DNA adducts)	DNA damage is a reflection of exposure and genotoxic effects	[129]
Triglyceride levels and IGF1	The levels of lipids (such as triglycerides) and of growth hormones (such as IGF1) can be used as a metric of the animals energy reserves	[129]
Biomarkers that integrate both chemical exposures and biological effects		
Acetylcholinesterase (AChE)	An enzyme that hydrolyzes neurotransmitters that is inhibited by exposure to carbamate and organophosphate pesticides	[99]
Tissue transcriptome	Microarray and RNA sequencing of tissue transcriptome can identify cellular pathways perturbed by chemical and environmental stressors	[130]

Also, oxidative stress biomarkers and other enzyme based biomarkers are known to exhibit a "bell shaped dose response curve" with increasing toxicant dose or exposure time [141]. This should be taken into account when designing field experiments. Furthermore, these parameters are significantly affected by nutritional status, age, and other non-chemical factors.

Another commonly used effects biomarker is the measurement of the integrity and stability of lysosomal membranes detected by the uptake and retention of a cationic dye, neutral red. Dose-response relationships have been found for a range of aquatic contaminants, including both metals and organics [142]. Lysosomal stability has been positively correlated with scope for growth, reproductive output, genotoxicity, total oxyradical scavenging capacity and protein synthesis and can be deemed as ecologically relevant [126]. Edge and colleagues [127], in a study of the Sydney rock oyster in contaminated estuaries containing both metals and PAHs, showed that lysosomal membrane stability was a more useful indicator of environmental stress than lipid peroxidation or concentrations of glutathione (GSH). Lysosomal stability correlated well with fertilisation, normal embryo development and estuary status.

Significant correlations were observed between lysosomal membrane stability and induction of the exposure marker MT in the digestive glands and gills and liposomal membrane stability, in mussels (*Mytilus galloprovincialis*) exposed to heavy meal contaminants [143], but it was noted that other factors such as age, size and seasonal variation could contribute to MT induction. The aforementioned findings and those of other studies support the use of multiple biomarkers covering both exposure and effects, rather than the use of a single indicator.

Genotoxicity is another important effect that is commonly measured using biomarkers. For instance, DNA integrity (measured as strand breaks with the Comet Assay) was inversely correlated with total oxyradical scavenging capacity in mussels collected from a eutrophic lagoon in coastal Italy [144]. Both a spatial and temporal gradient was measured, with mussels having poor metrics in the areas with the worst water quality and in the summer, when temperatures were elevated. Both metrics were correlated with a loss of biodiversity in the area [144]. DNA damage (measured using the Comet Assay) was also found to correlate with concentrations of sediment contaminants in mussels deployed in San Diego Harbour, CA., USA [145].

DNA damage can also be measured directly using DNA adducts. For instance, levels of DNA adducts were found to be elevated in fish captured near an aluminium smelter in western Norway [146]. Adduct levels correlated to both EROD (indicating PAH exposure) as well as skin ulcers and fin erosion [146]. DNA adducts were also used to measure the potential impacts of the Sea Empress oil spill [147]. DNA adducts were measured in a sponge, a mussel and several species of fish. Although elevated levels of adducts were not measured in either species of invertebrate, they were found in all three fish species. The return of these adduct to baseline levels after eighteen months indicated that fish populations were no longer under genotoxic threat [147].

2.5. Integrating Exposure and Effect Biomarkers

Incorporation of transcriptomic, or genome-enabled biomarkers, is becoming increasingly recognized as a powerful approach that can yield information on both exposure and pathways of injury (i.e., biological effects), and thus provide somewhat of a bridge between exposure and effects. Genomic approaches are widely accepted in the context of biomarker generation [148], however, these transcriptional approaches need to be validated with a phenotypic anchor, i.e., clearly defined and measurable effects at organ, tissue, or physiological levels. For example, whole tissue transcriptome profiling has been effectively used for screening adverse outcomes associated with drug and chemical toxicity in mammalian studies [149, 150]. Microarray approaches may be particularly useful in determining the contributions of various classes of environmental contaminants toward sub-lethal toxicity in mixture scenarios [151, 152, 153].

Increasingly, microarray-based gene-expression profiling, or other methodologies that measure overall gene expression, are being used as "discovery" tools to characterize response in both laboratory and field studies. These profiling techniques have the advantage of being able to be used when the stressors are unknown and being able to identify the causative agent of decline in the organism's health, since any field results can be compared to results from controlled laboratory studies. They are also inherently multivariate, in that transcript levels of biomarkers can be assessed simultaneously with the transcripts for gene products more closely associated with fitness. The disadvantage of these techniques is that interpreting the data can be challenging, especially given that the No Observable Transcriptional Effects Level (NOTEL) is often far below the concentrations at which toxic effects are seen [154, 155], and that the annotation of many genomes is incomplete.

Van Straalen and Feder [156] considered transcriptomics as a 'super-biomarker' because of the wealth of information generated but noted the current difficulty in linking the measured gene expressions to ecologically relevant effects.

RNASeq is based on measuring the abundance of transcripts, but unlike microarray, is not limited to commercially available array chips or customising expensive individual chips, and does not require pre-existing knowledge of the fish genome [157]. This means that RNASeq can be used to assess responses to contaminants in non-model fish species [157]. However, the application of this biomarker needs to be further development to provide clear connections between external levels of contaminant exposure, changes in transcripts abundances, and early adverse effects in organisms.

2.6. Biomarkers of Susceptibility

Biomarkers of susceptibility indicate the inherent or acquired ability of an organism to respond to the challenge of exposure to a specific xenobiotic substance, including genetic factors and changes in receptors, which alter the susceptibility of an organism to that exposure [158]. Furthermore, they help to elucidate variations in the degree of responses to toxicant exposure observed between different individuals. Some biomarkers can also indicate susceptibility to adverse outcomes from environmental contaminants, although these have not been developed or incorporated in ecological assessment frameworks due to the fact that they largely rely on well-defined genetic databases and the incorporation of epidemiological studies, which are areas of future development in the field of aquatic toxicology.

Evidences indicate that inherited and acquired genetic susceptibility, epigenetic modifications as well as alterations in physiological structures and functions induced by age, pathological conditions, and lifestyle factors, may lead to different phenotypic expressions from xenobiotic exposures. Particularly, inherited genetic susceptibility may play a role in influencing the individual response to exogenous exposures in a complex "gene–environment" interaction [159].

Therefore, understanding which genetic polymorphisms, genotoxic changes, epigenetic profiles and host factors may affect the toxicokinetic and dynamic nanoparticle (NP) modelling, appear essential to get insights into the still not understood NP exposure/disease continuum and to identify susceptibility biomarkers indicative of an elevated sensitivity to NP effects. This seems an even more challenging issue considering that the same great variability in NP physicochemical properties, i.e., in terms of size, chemical composition and surface area, that make them so attractive for a variety of product applications may also prove complex and changeable exposure scenarios, potentially influencing individual response to NP toxicity. These biological indicators may be useful to provide quantitative estimates of a population variability to be employed into an adequate occupational NP risk assessment and consequently in the plan of specific or implemented workplace preventive and protective measures.

Chapter 3

MOLECULAR EVOLUTION

Biomarkers can reflect the actual state of health of organisms or populations, but do not provide information on the potential of a population to survive in a polluted environment. Genetic approaches can, however, produce relevant knowledge in an ecotoxicology study. Indeed, the genetic variation in populations results from their evolutionary histories and represents the raw material that has, or has not, enabled the evolutionary adaptation of these populations faced with environmental challenges (global change, hypoxia, emerging contaminants, multistress, etc.).

Genetic tools can be used particularly to analyze the differential capacity of organisms (i.e., of genotypes) in a population to cope with a chemical stress, although the between individuals variability in biochemical or physiological responses is still being widely ignored in ecotoxicology, or is considered background noise [160].

Genetic investigations in ecotoxicology can actually lead to:

1. The assessment of within-population genetic diversity, a key factor allowing species survival; any alteration of this diversity could lead to a possible loss of adaptive capacity to meet any additional stress
2. The estimation of the potential connectivity between populations by the analysis of the between-populations genetic diversity. Are the

populations more or less isolated and thus more or less genetically differentiated? What is the level of gene flow between populations? Does this level allow a possible recolonization of the habitat by a species, after a massive demographic bottleneck?

3. The identification of relationships between genetic diversity and physiology by searching for possible couplings between genotypes and phenotypes, considering fitness-relevant biological traits.
4. The detection of modifications at gene expression level potentially leading to cellular damage, or to physiological acclimation or even to resistance to chemical stress.

Which forces are responsible for changing genetic variability? Across the whole distribution area of a particular species, random mating between individuals is indeed impossible, the dispersion–migration capacity of the species being limited. Thus, one observes generally in the field different local populations (or demes) considered as reproduction units displaying spatial and temporal stability. Within-population genetic variability is subjected to four major evolutionary forces (mutation, genetic drift, migration, and selection) (Figure 6) that can lead to changes in allele frequencies [161, 162].

3.1. Mutation

Mutation is a process that creates genetic diversity, producing a new allele different from the ancestral allele. The different alleles observed for a particular gene are derived from mutational modifications frequently linked to errors during DNA replication. Mutations are rare events (single locus mutation rate per generation: 10^{-4} to 10^{-6}); thus, they can change allele frequencies in a population, but these changes are usually very slow.

Mutations in noncoding regions generally have no impact on the phenotypes and thus are not submitted to selection; this kind of polymorphism is particularly analyzed in population genetics to characterize within- and between-population genetic diversity.

Mutations in coding sectors for a functional protein or for a transcription factor can considerably affect individual physiological performance or organism fitness.

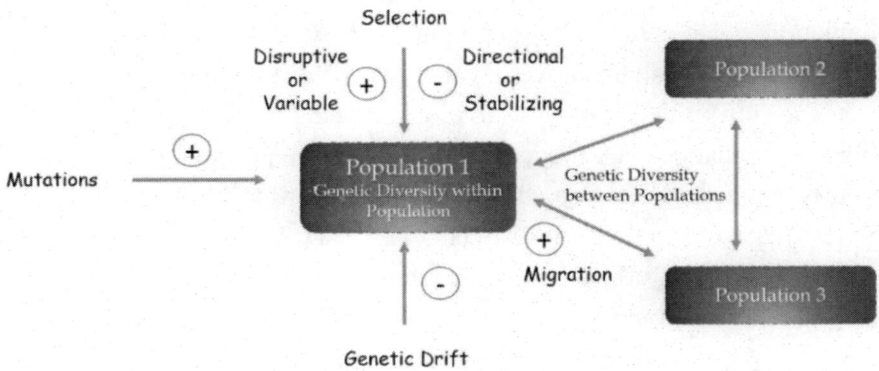

Figure 6. Effects of the four main evolutionary forces on within-population genetic diversity: a process leading to conservation or increase +, or to reduction – of the within-population genetic diversity.

Particular noncoding regions of the genome consist of many copies of the same sequence (two to nine nucleotides) and are called microsatellites. These markers display a high natural mutation rate, and thus are particularly targeted in explorations of the genetic structure of populations in the field and to analyze mutation load in populations.

3.2. Genetic Drift

Genetic drift refers to fluctuations in allele frequency that occur by chance, especially in small populations, as a result of sampling error. In small populations, in particular, genetic drift effects are frequently relatively dominant to those of selection, high-performance genotypes and genotypes sensitive to stress displaying the same extinction probability [163]. Genetic drift leads to an increase in the genetic differentiation between populations.

3.3. Migration

Migration is characterized by the movement of individuals between populations, whereas gene flow is the movement of genes between populations, i.e., is linked to the successful reproduction of an immigrant in its new habitat. Generally, gene flow leads to an increase of within-population genetic diversity and to a decrease of between-populations genetic diversity, i.e., to a lower level of genetic differentiation.

Gene flow between moderately contaminated populations and heavily polluted populations can induce a reduction of the adaptation speed in the polluted systems [164]. In a contaminated population, the identification of immigrants by genetic tools can be particularly pertinent, allowing the exploration of the effects of contaminants, primarily on residents and secondly on immigrants [165].

3.4. Selection

Selection preferentially retains in a particular environment the individuals displaying the higher fitness. Selection can oppose genetic drift in extreme environments, where selection can maintain a particular polymorphism over thousands of generations, even for isolated populations of a few dozens of individuals [166, 167]. The demonstration of the effects of selection on molecular variants (i.e., the estimation of their values in terms of physiological performance) is generally very difficult to carry out, DNA diversity being very complex to analyze. Thus, selective pressure is sometimes more easily detected by measuring the heritability of a phenotypic trait in a population, considering that the trait is integrating the physiological effects of several genes [163].

The selection modes induced by contamination on populations are diversified and complex [168]. If a polymorphism is linked to a pattern of resistance, one should detect an increase in the tolerant genotype frequency, possibly by different modes:

- Directional selection, without a heterosis effect (i.e., without heterozygote superiority), where a reduced genetic variability is observed over the population.
- Stabilizing selection, fitness being highest at some intermediate phenotypes and lower at the phenotypic extremes (decrease of genetic variability).
- Variable selection in space and time (alternately toxic and nontoxic periods); if fitness is highest for the tolerant genotype in the toxic period but lower in the nontoxic period, genetic variability will thus be maintained.
- Disruptive selection where fitness is lowest at some intermediate phenotype and is higher at the phenotypic extremes; genetic diversity will be also maintained.

3.4.1. Markers of Genetic Diversity in Ecotoxicology

Effects of contaminants on genetic diversity can be considered indirectly through the measure of biometric characters linked to fitness with a heritable component (quantitative genetics approach), or directly through the study of allozymes or DNA markers (population genetics approach).

Biometric characters, potentially under control of several genes, may generate differential physiological performances and thus lead to population effects in the face of environmental stress [169]. These biometric approaches are based on the hypothesis that selective pressures determine the genetic characteristics of a population and thus its adaptive potential. For these approaches, large laboratory experiments performed on successive generations are essential and usually has to be studied the genetic component of the physiological trait associated with tolerance (i.e., its heritability).

The second family of markers associated with genetic diversity concerns molecular markers. Allozymes (enzymatic proteins) were the first markers to be used in populations facing environmental stress. Allozyme loci may have several allelic variants with different biochemical characteristics, in particular, molecular variations that induce variations in charge or structure, leading therefore to differences in migration speed during electrophoresis.

Considering several individuals and several allozyme loci, the number and the frequency of the different alleles of the whole population may be estimated and its genetic diversity characterized. Enzymatic polymorphism reveals about 30% of the DNA genetic variability since many molecular mutations do not lead to variations in charge or structure of the subsequent protein.

The “tolerance to contaminants/genotype” association of allozyme loci has been encountered in many aquatic organisms since the 1980s, both in freshwater and marine organisms. For some authors, allelic variants are important for the determination of tolerance [170]. For others, heterozygosity of particular loci [171] or multiloci heterozygosity [172] are determinant characters for tolerance.

Allozyme markers can still be considered pertinent tools for the evaluation of genetic variability in ecotoxicology, but more often, current research considers genetic variation directly at the DNA level, which represent the genomic approaches. At the DNA level, two types of markers—neutral markers and markers under selection— are important to consider in ecotoxicology.

Neutral markers are loci not evolving directly in response to selection [167], the dynamics of which are controlled mainly by mutation, genetic drift, and migration. These markers are usually noncoding regions of the nuclear DNA (such as microsatellites, introns, and intergenic regions) and mitochondrial loci (the control region called the D-loop, the different subunits of NADH, etc.) and are useful in ecotoxicology:

- For identifying genetic structure in natural populations and, in particular, potential gene flow between chronically contaminated populations and reference populations
- For exploring the genetic history of populations in different environmental contexts
- For detecting the impact of drastic reduction in population size that can lead to the loss of genetic variation through genetic drift

Their level of variability may, however, be affected by mutagen toxicants or influenced by selection on nearby (linked) loci (through the phenomenon of hitchhiking).

In an ecotoxicological context, markers under selection are fragments of gene that code for a protein conferring a selective advantage or disadvantage [167] on the individual subjected to a chemical stress. Identifying the genes that potentially affect particular traits is the major difficulty. Studies were initially focused on individual candidate genes such as the gene coding for xenobiotic metabolizing enzymes (i.e., cytochrome P450s, glutathione-Stransferases), for a transcription factor AhR (aryl hydrocarbon receptor), a well-known chemical sensor, or for proteins involved in energy production [173, 174, 175].

The identification of their polymorphism relied in particular on techniques such as restriction fragment length polymorphism (RFLP), single strand conformation polymorphism, and denaturating gel gradient electrophoresis [167].

Recent advances in molecular technologies have made genome-wide sampling in populations feasible through genome typing [176]. Genome typing is the simultaneous genotyping of hundreds of loci from across the genome [microsatellites, amplified fragment length polymorphism (AFLPs), or nucleotide polymorphisms], combining both neutral markers and markers potentially under selection [167].

The classical AFLP protocol uses a polymerase chain reaction PCR-based DNA fingerprinting technique to produce dozens of polymorphic markers (presence or absence of a restriction enzyme site), that cover the entire genome. AFLPs are increasingly used to identify markers associated with traits under selection in non-model organisms [177, 178]; these markers were successfully used to detect signatures of selection in natural populations adapted to chronic pollution [179].

The Diversity Array technology (DArt) approach uses a microarray, in which each spot contains a DNA fragment that has been amplified from a library of polymorphic markers that were identified during an initial screening phase [180]. DArt is attractive because a single PCR can amplify hundreds of polymorphic markers and because automation is easier using images rather than gel electrophoresis. However, this technique can only be used on species for which large data sets and effective genomic tools have been developed.

Single nucleotide polymorphisms (SNPs) have recently attracted the interest of geneticists and ecotoxicologists; they require species-specific marker development, but they have potential for automation and high throughput [181]. The rapid development of numerous SNPs (including nonsynonymous and functional SNPs) is becoming feasible in some nonmodel species, owing to the rapid growth of expressed sequence tag (EST) databases, data-mining software, and primer-design strategies such as Comparative Anchor-Tag Sequences and Exon-Primed Intron-Crossing PCR [182, 183].

3.5. Establishing Links between Chemical Stress, Genetic Diversity, and Phenotypic Responses

Interest in evolutionary ecotoxicology (i.e., the study of effects of chemical pollutants on the genetics of natural populations) is rapidly growing among ecotoxicologists; the major population genetic impacts in the context of contamination studies include (1) changes in allelic or genotypic frequencies caused by increased mutation rates, (2) genome-wide changes in genetic diversity linked to demographic bottlenecks, (3) changes in allelic or genotypic frequencies caused by contaminant-induced selection, and (4) changes in dispersal patterns of gene flow that alter the genetic relationship among populations [184].

Evidence of direct effects of pollution on genomes has been highlighted in terrestrial organisms and especially in the case of radionuclide contamination. For example, Dubrova and colleagues [185] found that germinal mutations were two times higher at three minisatellite loci in a human population living around Chernobyl compared to a reference population living in England. Ellegren and colleagues [186] highlighted higher germinal mutations in swallows (*Hirundo rustica*), whose parents had reproduced in the radionuclide-contaminated zones of the Chernobyl region compared to reference swallows; they noted higher germinal mutations not only in noncoding regions of the DNA (especially microsatellites), but also in coding regions such as mutations causing albinism.

Herring gulls (*Larus argentatus*) nesting near steel mills on the Great Lakes were shown to have higher germline mutation rates at minisatellite DNA loci than those at rural sites [187], and mutation frequency increased with colony proximity to integrated steel mills [188]. The authors postulated that inhaled airborne contaminants emitted from steel mills, such as polycyclic aromatic compounds, were largely responsible for mutation induction [188]. In order to address this issue, mice were exposed in situ to ambient air in a polluted industrial area near steel mills. Heritable mutation frequency at tandem-repeat DNA loci in mice exposed 1 km downwind from two integrated steel mills was 1.5- to 2-fold elevated compared with those at a reference site 30 km away [189].

Indirect effects of pollutants on genomes may be due to a massive reduction of the size of the population due to mortality or a selection of alleles or genotypes associated with "tolerance" in contaminated areas and an elimination of the most "sensitive" genotypes.

Only a few studies have highlighted the existence of genetic erosion due to a massive reduction of the size of populations in contaminated sites. For example, Demarais and colleagues [190] concluded that the modifications of genetic structure of populations of the freshwater fish (*Gila seminuda*) observed after an acute accidental exposure to the insecticide rotenone, were probably due to a genetic bottleneck caused by a high mortality. Murdoch and Herbert suggested the same hypothesis to explain the weak mitochondrial genetic diversity observed in populations of catfish (*Ameiurus nebulosus*) from contaminated sites. Facemire and colleagues [191] have also hypothesized that the mixture of contaminants [mercury, polychlorobiphenyls (PCBs), etc.] occurring in Florida induced mortalities in Florida panther (*Puma concolor*) populations, leading to a loss of genetic diversity in these populations. In a more recent study, Matson and colleagues

[192] observed a general reduction of the haplotype and nucleotide diversities in marsh frog (*Rana ridibunda*) populations in the highly contaminated industrial sector of Sumgayit, Azerbaijan, compared to uncontaminated reference sites.

Many studies conducted in the past three decades, together with the previously cited studies, have suggested that contamination may constitute a strong selective pressure, able to lead to adaptive changes in natural populations. However, the specific identification of the target on which selection acts remains very explorative, even in the case of the mummichog (*Fundulus heteroclitus*), the premier teleost fish model in environmental biology, for which the genetic basis for resistance to pollution still needs to be elucidated [175, 179, 181].

Thorough analyses of the molecular mechanisms of resistance to toxicants were first performed during experiments conducted with antibiotics, medicine, and pesticides [193]. Several modifications of the target were detected:

- Punctual mutations in the coding regions of the enzyme targeted, leading to structural alterations of the active site and thus to a weaker sensitivity of the enzyme
- Duplications of the gene, thus leading to more efficient expression
- Mutations in the regulatory region of a gene (promoter) or in a gene coding for a protein that binds to a promoter

In the case of complex contaminations, such as those observed in aquatic systems, mechanisms of selection remain, for the most part, unknown. Selection may possibly act on loci physically close to those studied. Proximity on a chromosome may allow genes to be dragged along, through a mechanism of "hitchhiking," with a selective sweep experienced by the advantageous gene nearby [194].

3.6. Studies about Molecular Evolution of Biomarkers

In the following section will be presented studies examples related to the molecular evolution of several molecular biomarkers of exposure and effects. All these studies were performed by Rigers Bakiu and colleagues, where most of the bioinformatics work was done by him personally.

3.6.1. P-Glycoproteins

Many aquatic species are able to survive in environments, which contain high levels of multiple anthropogenic pollutants or natural product toxins. Overexpression of a 170 kDa transmembrane P-gps was found to prevent the accumulation of cytotoxic drugs in resistant cells. P-gp is found endogenously in specialized epithelial tissues involved in secretion and excretion such as the mammalian gut, liver, and kidney, as well as on endothelial cells of capillary blood vessels at the blood-brain barrier. P-gp acts as an energy-dependent pump to translocate a wide variety of structurally and functionally diverse substrates. These compounds tend to be moderately hydrophobic, planar, natural products, which are often substrates for or metabolites of detoxification enzymes, such as cytochromes P450 (CYPs) [195]. P-gps prevent the cellular accumulation of endogenous metabolites, phospholipids, and xenobiotics in exposed animals and cell cultures. P-gp-like proteins have been described in a variety of aquatic organisms including sponges, mussels, oysters, clams, worms, and fish.

As a specific transmembrane efflux transporter that pumps structurally different xenobiotics out of the cell, in Pgp (Abcb1) appeared to be the key mediator of the so-called MXR defence system [196, 197] in aquatic organisms. It has been shown that the MXR phenomenon is constitutive to aquatic organisms, inducible in response to pollution [198, 199, 200], and sensitive to specific environmental pollutants/chemicals commonly called MXR inhibitors or chemosensitisers [201, 202].

Both natural products and anthropogenic contaminants found in the aquatic environment appear to be substrates and inducers of the multixenobiotic resistance transporter in aquatic organisms. These observations suggest that in addition to normal cell function, P-gp activity may contribute to the relative hardiness of some aquatic species exposed to xenobiotics. In addition to well characterized detoxification systems (phase I, II, III enzymes, heat shock proteins, etc.), the induction of a multixenobiotic defense mechanism in organisms living in polluted environments may explain why contaminant spills cause more severe adverse effects at pristine sites than in already polluted areas [195].

P-gp, member of the adenosine triphosphate (ATP)–binding cassette (ABC) superfamily [203], has unusually broad poly-specificity, recognizing hundreds of compounds as small as 330 daltons up to 4000 daltons [204, 205].

Most P-gp substrates are hydrophobic and partition into the lipid bilayer [206, 207]. Thus, P-gp has been likened to a molecular "hydrophobic vacuum cleaner" [208], pulling substrates from the membrane and expelling them to promote MDR and MXR.

The X-ray structure of apo P-gp at a resolution of 3.8 Åwas recently obtained [209] (Figure 7). It revealed distinct drugbinding sites in the internal cavity capable of stereoselectivity, which is based on hydrophobic and aromatic interactions. Apo P-gp structures have an inverted "V" shape, inward-facing conformation, for drug entry, whereas the outward-facing conformation releases the substrate to the extracellular medium [209].

Expression of MXR proteins may be an ecotoxicologically important characteristic, as it could critically influence the susceptibility of aquatic organisms to pollutants [210, 211]. The tissue distribution of P-gps is indicative of their function in accelerating excretion or preventing the uptake of xenobiotics. In addition to their occurrence in tumor cells [212, 213], expression of P-gps associated with an mdr1-like mechanism have been detected in normal teleost tissues, that are involved in a secretory, absorption or barrier function such as the liver, kidneys, gills and intestine [214, 215, 216].

Teleost fishes, with about 27 000 species [217], are the largest and most diverse group of vertebrates. Teleosts account for more than 99% of ray-finned fishes (*Actinopterygians*), which diverged from lobe-finned fishes (*Sarcopterygians*) about 420 million years ago (Mya). They exhibit remarkable diversity in their morphology, behavior, and adaptations [218]. These evidences prompted Bakiu [219] to investigate the molecular evolution of P-gp in such a high number of candidate species taxonomy group. For this purpose, several phylogenetic analysis were performed using teleost amino acid and cDNA sequences.

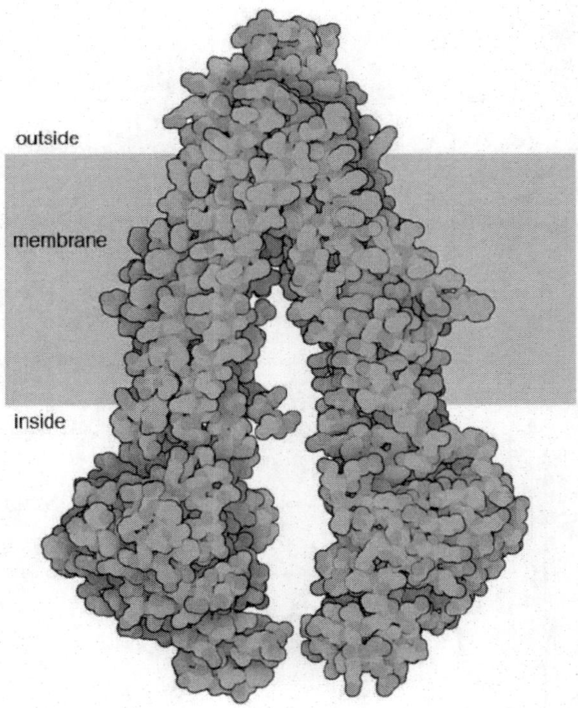

Figure 7. Mouse P-gp 3-dimensional structure (PDBID: 3G60).

From the comparison of the analyzed species P-gps with the actinopterygian time-calibrated phylogeny based on nine nuclear genes and 36 fossil age constraints [220], from the bioinformatics analyses emerged a good correspondence between the phylogenies. In the actinopterygian phylogeny, cypriniformes and salmoniformes were positioned close to each other and far from perciformes and other *Acanthopterygii* superorder members.

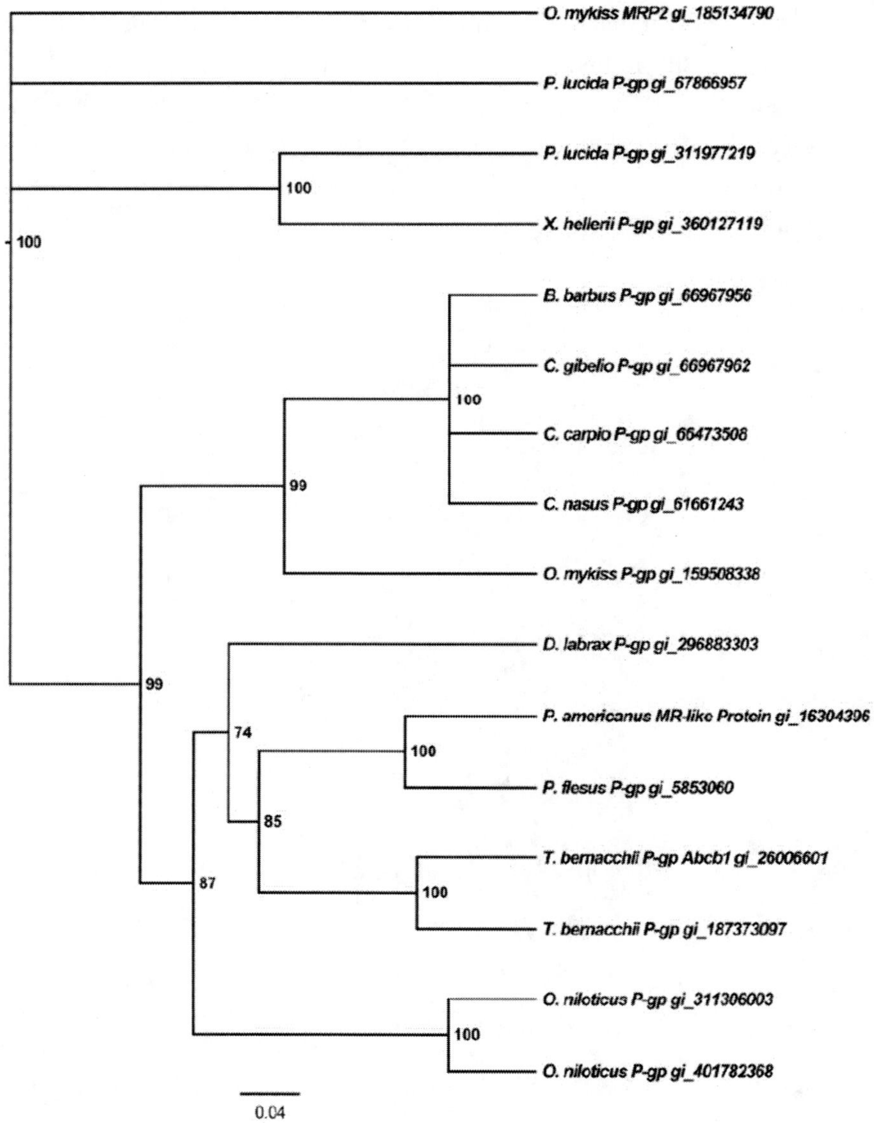

Figure 8. Phylogenetic relationships among P-gp cDNA sequences using BI method (arithmetic mean = -21757.714; harmonic mean = -21777.237). Posterior probability values higher than 50% are indicated on each node. The scale for branch length (0.04 substitution/site) is shown below the tree.

Teleosts P-gp phylogeny (Figure 8) showed an identical topology to actinopterygian phylogeny except for European flounder (*Platichthys flesus*) P-gp. Salmoniformes and cypriniformes P-gps were positioned closer to each other than to Acanthopterygii superorder members. The good correspondence between P-gp and teleosts phylogeny suggest that P-gp molecular evolution had followed the same evolution pathway as the speciation of the analyzed teleost species.

Clearfin livebearer (*Poeciliopsis lucida*) and European flounder P-gp genes could be good examples in teleosts demonstrating active process that included adjacent duplications, presumably by unequal crossovers; nonadjacent duplications believed to occur by chromosome or genome duplication; and the loss of genes through deletions and sequence degeneration, observed in zebrafish (*Danio rerio*) ABC genes [221]. However, further analyses are needed to confirm previous hypothesis. Generally, P-gp molecular evolution in teleosts seemed to follow identical evolution pathways to teleost fishes speciation. Other teleosts P-gp cDNA and amino acid sequences would be extremely useful in P-gp molecular evolution studies. Evolution studies could give further helpful information for fish P-gp structural and functional studies

3.6.2. Metallothioneins

Metalothioneins (MTs) are cysteine rich peptides occurring mainly in the cytosol and in the nucleus and lysosomes. They are nonenzymatic proteins with a low molecular weight, high cysteine content, no aromatic amino acids and heat stability. The thiol groups (–SH) of cysteine residues enable MTs to bind particular heavy metals. MT-like proteins have been reported in many vertebrates including many species of fish [222, 223] and aquatic invertebrates mainly molluscs [224] and crustaceans.

MTs can be induced by the essential metals Cu and Zn and the non-essential metals Cd, Ag and Hg in both vertebrates and invertebrates. Thus the exposure of marine organisms to toxic metals can lead to changes in several biochemical processes that have the potential to be used as biomarkers of exposure and therefore as 'early warning' signals of the presence of these particular contaminants. The induction of MTs was reported in the oyster (*Crassostrea virginica*) [225] and the mussel (*Mytilus galloprovincialis*) [226]. About 50 different species of aquatic invertebrates, the majority of which are mollusks or crustaceans are found to show response to induction of MTs. Thus they are used for evaluation of pollution in the marine environment and are seen as potential biomarkers of metal exposure in mollusks and fish [227]. Tissues directly involved in metal uptake, storage and excretion have a high capacity to synthesize MTs. In aquatic organisms, these proteins have been identified in the digestive gland (also termed the midgut gland or hepatopancreas) and gills of molluscs and crustaceans [228, 229].

The role(s) of MTs in the physiological processes can be explained by the fact that it is primarily involved in essential and non-essential metal pathways. The activities of MT can be observed in elements of both homeostasis and detoxification. MT binds to excess of essential or pollutant metals. Thus it protects the organism against toxicity by restricting the availability of these cations at detrimental sites. Some toxic metals such as Ag, Cd, Cu, Hg and Zn have a high binding affinity for cysteine. In some cases, MTs have high cysteine content (30%), low molecular weight, heat-stability, and a strong affinity for binding metals. In the case of mollusks MT have a high glycine content differentiating them from most of the other MT isoforms known today [223].

Anyway, MTs are distinctively identified in only a few species of marine molluscs (*Patella vulgata*, *Crassostrea virginica*, *Patella granularis*, *Mytilus edulis*) [230, 231] and recently in *Ruditapes decussates* [232] and several fish species [233].

The induction of MT has been detected in organisms from contaminated areas or *in vitro* experiments with exposure to metals such as Ag, Cd, Cu, Hg and Zn in laboratory. The extent of MT induction can vary between species and between tissues. The sequestration of metals by MT is clearly evident in the gills, digestive gland and kidney, indicating the significance of these tissues in uptake, storage and elimination of metals [234]. The use of MT as biomarker has been validated in many *in situ* studies [235, 236,

237, 238]. The results are generally positive, in particular when the metal gradient of pollution is quite real. More and more of studies *in situ* combine the quantification of several biomarkers, of which MT is one them [239, 240, 241, 242, 243, 244].

Numerous studies demonstrate Cd induced MT expression in a variety of fish species including the turbot [245], carp [246], goldfish [247], sole [248], zebrafish [249], rainbow trout [250] and tilapia [251]. MT transcript and protein are induced in fish by other bivalent metals, including Zn, Cu, Pb and Hg [252, 253, 254, 255, 256, 257]. Therefore, MTs have become of great interest for assessing pollution in the marine environment and are seen as potential biomarkers of metal exposure in fish [258, 259, 260, 261]

Gilthead seabream (*Sparus aurata*) is one sentinel fish, which is used by several authors due to its widely distribution throughout the entire Mediterranean Sea [262, 263]. *S. aurata* MT has a molecular weight of 5.966 kDa, contains 20 metalbinding cysteine residues and has already been proposed as a biomarker of heavy metal exposure [264]. Previous studies results [265] suggest that there is considerable interaction among MTand glucose-6-phosphate dehydrogenase (G6PDH) induction by metals in *Sparus* sp. (*Sparus sarba*). Both the Cd-induced MT and G6PDH responses may work together to protect cells against oxidative damage, since MT and GSH (generated by the G6PDH reaction) are both cellular thiols capable of protectin cells from Cd toxicity by virtue of their ability to sequester metals. Indeed, Schlenk and Rice [266] clearly demonstrated that concomitant induction of MT and GSH by metals may provide better protection against oxidative damage than MT induction alone.

Ghedira and colleagues [267] determined the value of MT induction using the assay of Viarengo [268] and cadmium and copper accumulation in liver, gills and kidney of *S. aurata* after short term exposure (48 hours). They noted that MT level induction related to the higher metal accumulation in three organs revealed the main role of MT in metal homeostasis and detoxification in *S. aurata*. Similar results were previously reported by Filipovic and Raspor [269], who found a high correlation between Zn and Cu accumulation and MT induction in liver of M. surmuletus and L. aurata.

For farmed fish, growing conditions (food and water chemistry) may determine metal composition of fish tissues as well as the response that these fish exhibit to metal toxicity. For this reason, Cretì and colleagues [270] examined the concentration of the heavy metals, Cd and Pb, in tissues of gilthead seabream collected from different types of fish farming.

Heavy metal concentrations in feed used in fish farming were measured to determine the principal source of metals in fish tissues. In addition, the relationship between metal concentration and MT content in tissues was investigated. Other studies of heavy metal concentrations in tissues from different types of breeding allowed them to identify and define the potential impacts of the different aquaculture systems on the welfare of farmed fish. Data relating to muscle tissues demonstrate that in general, metal concentrations were lower in muscle compared to gills, liver, gut and kidney. This is an important fact, because muscle constitutes the greatest mass of the seafood that is consumed, and this result was in good agreement with observations carried out on several wild and cultured fish species [271, 272, 273, 274, 275].

Furthermore, it was noteworthy the fact that in the fish tissues, the content of both Cd and Pb was not strictly correlated with that of metallothionein. Indeed, the marked accumulation of both metals in liver, as well as the high Pb content in gills and kidney, was not accompanied by a concomitant accumulation of MT in these tissues. The results clearly demonstrated that MT levels were also tissue-specific, with the highest levels in the muscle, followed by gut, liver, kidney and gills. However, the absence of a correlation between MT content and metal accumulation in tissues has been already reported for many biological systems [276, 277, 278]. Trinchella and colleagues [277] have postulated that metal ions are mostly present in tissues not as free ions but possibly as metal–protein complexes, so the amount of free ions could be too low to bring about a measurable induction of MT expression and synthesis. The high MT level found in muscle could be explained considering that this tissue is the site of an intense oxidative metabolism [279] with the consequent production of a high amount of free radicals which, in turn, could induce the MT synthesis [280].

Another important consideration is that the highest MT content in the muscle of fish breed in the extensive system was in agreement with the high metabolic rate of these fish that, as stated above, could be considered as wild fish. The high MT level found in the muscle of fish from the intensive breeding could be associated with the high stocking density generally present in this type of breeding, which is the main cause of stress for fish.

All the information presented in Bakiu and colleague's review [281] could help us to further understand the precious role of metallothioneins as biomarker in ecotoxicological studies and suggest one of scientific approach to use in ecosystem biomonitoring programs in order to have a healthy ecosystem with healthy beings like humans. Using metallothionein as marine pollution biomarker could be extremely helpful to improve fish productivity and the aquaculture production system quality.

In relation to molecular evolution considerations, recently Nina Seren and colleagues [282] did a detailed analyses on molecular evolution and revealed functional divergences of the metallothionein gene family in vertebrates. Phylogenetic and synteny analyses indicated the existence of a eutherian MT1 and MT2, a tet-rapod MT3, an amniote MT4, and fish MT. The optimalgene-tree/species-tree reconciliation analyses identified the best root in the fish clade. Functional analyses reveal vari-ation in hydropathic index among protein domains, likelycorrelated with their distinct flexibility and metal affinity. Furthermore, the analyses of functional divergence identified amino acid sites correlated with functional divergence among MTs. Uncovering the number of genes and sites possibly corre-lated with functional divergence and it would help to design cost-effective MT functional and gene expression studies.

The dataset used in Nina Seren and colleagues [282] paper was based on representative vertebrate species with complete genome annotations that could be confirmed by more than one genomic database to improve data quality.

This approach improves detections of ortholog and paralog genes, improving resolution of the molecular evolution of the MT gene family in vertebrates. Using this dataset, Nina Seren and colleagues [282] were able to recover multiple MT types in all amniotes, suggesting that duplication and functional divergence in MTs is not limited to mammals and birds. Furthermore, our results Nina Seren and colleagues [282] suggest the existence of an amniote MT4 clade, a mammalian MT1 and MT2 clade, and a tetrapod MT3 clade. All these results, together with the analyses of functional divergence between mainMT clades, suggest a likely association between MT functional divergence and duplication events in vertebrates and a marked functional distinction between MT1 and MT2 *versus* MT3 and MT4 in vertebrates.

In humans, MT1 and MT2 are inducible and expressed in almost every tissue. MT3 and MT4 are, on the other hand, relatively unresponsive to inducers that stimulate MT1 and MT2 expression and are mostly located in the central nervous system and in the stratified squamous epithelium, respectively [283]. The limited data available on the expression and induction of distinct MT types in non-mammalian vertebrates and non-vertebrate chordates, e.g., [284, 285], which suggest the existence of two MT types in non-mammalian vertebrates, one that is more ubiquitous, and the other that is more specialized.

Though the poor resolution of phylogenetic relationships among the main MT types does not allow a full interpretation of the evolutionary process of functional divergence in this gene family and to infer whether a more generalized MT type evolved into one or more functionally specialized MT types in amniotes. More biochemical and expression data are needed to understand the underlying mechanisms of functional divergence after gene duplication in MTs, especially between mammals and other vertebrates. Our results on the number of ortholog genes occurring in each studied species and the sites potentially involved in functional divergence among MT types can help design future MT functional studies in other vertebrates, besides humans and mice. In fact, while a large body of bio-chemical and molecular work is currently available for mammalian model species, similar data are currently lacking for the distinct MT types recovered in non-mammalian vertebrates. Furthermore, in vertebrates, MT expression and concentrations are often used in toxicological and metal homeostasis studies [286, 287, 288]. As studies of mammalian model species reveal, not all MT types are equally involved in the same function or expressed in the same tissue and at the same time, nor do they show the same metal affinity. Therefore, the lack of knowledge on similar potential differences among MT types in other vertebrates possessing multiple MT genes may be misleading or provide incomplete conclusions. Comparative genomic and biochemical studies will help fill this knowledge gap and contribute to understanding of both MT evolution and functional divergence following gene duplication in vertebrates.

3.6.3. Heat Shock Proteins

Heat shock proteins consist of several families of highly conserved proteins that play an essential role in a number of cellular processes. Among the 70 kDa family of heat shock proteins, heat shock cognate protein 70 kDa (Hsc70) and inducible heat shock protein 70 kDa (Hsp70) have been most extensively studied in mammals [289, 290]. As other members of heat shock protein family, Hsp70 and Hsc70 have beenshown to be responsive to a wide variety of physiological and environmental insults, including thermal shock, heavy metals, free radicals and microbial infection.

Following stress stimuli and under normal physiological conditions, a major function of them is that of an intracellular molecular chaperone supporting protein folding and transport [291], and they are required for growth, survival and apoptosis of cells [292, 293]. In addition to serving as molecular chaperones, evidence is accumulating that mammalian Hsp70 and Hsc70 have been implicated in both innate and adaptive immunity [294, 295]

According to different authors, both Hsp70 and Hsc70 play important roles in antigen presentation, T-cell receptor complex formation [296], autoimmunity and tumor immunity [297, 298, 299]. In particular, immunomodulatory functions of extracellularly located or membrane-bound Hsp70 have been established [300, 301] since the extracellular Hsp70 can be recognized by immune cells including macrophages, dendritic cells and natural killer cells [302, 303].

Actually, it has been suggested that Hsp70may play dual role as chaperone and cytokine in mammalian immunity [304]. There is of great concern the monitoring of the impact of aquatic environments on fish. In view of their responsiveness to diverse forms of stress, Hsp70 or Hsc70 could be used as a sensitive biomarker for different classes of environmental assault [305]. Based on the considerations mentioned above, the identification and characterization of teleost Hsp70 and Hsc70 have received increasing attention in recent years. To date, the cDNA sequences of both Hsp70 and Hsc70 have been isolated from several fish species including seabream [306], rainbow trout [307] and Wuchang bream [308]. More efforts are being made to examine the expression of Hsp70/Hsc70 under different stimuli stress, including thermal effect [309, 310], environmental pollution [311], heavy metals [312, 313], insecticides [314] and crowding stress [308].

As in mammals, both Hsp70 and Hsc70 are induced in the pattern of gene expression under stressful conditions,(it is suggested) that they may be used as indicators of cellular stress response, when fish is exposed to environmental contaminants. Recently, a few studies reported that mRNA levels of Hsp70 were increased in bacterially infected fish [313, 315], providing valuable knowledge on the involvement of Hsp70 in fish immunity. Although Hsp70 is among the most studied genes at the molecular and cellular level, its molecular evolution is not well understood in fish. For these reasons, tt would be worth to pay a particular attention to the Hsp70 molecular evolution of percoid fish species, which are known to be very important in aquaculture and recreational fishery. The results of our phylogenetic analyses suggested that the analyzed organisms Hsp70 molecular evolution didn't respect organism phylogeny. The comparison of the phylogenetic trees showed that the best resolved phylogenetic tree was the amino acid sequences based phylogenetic tree (Figure 9). The nodes of this phylogenetic tree were supported by higher baesian posterior probability values than the nodes of coding cDNA sequences based phylogenetic tree. Generally, the cDNA based phylogenetic trees are more resolved than the amino acid based phylogenetic trees, but this was not that case.

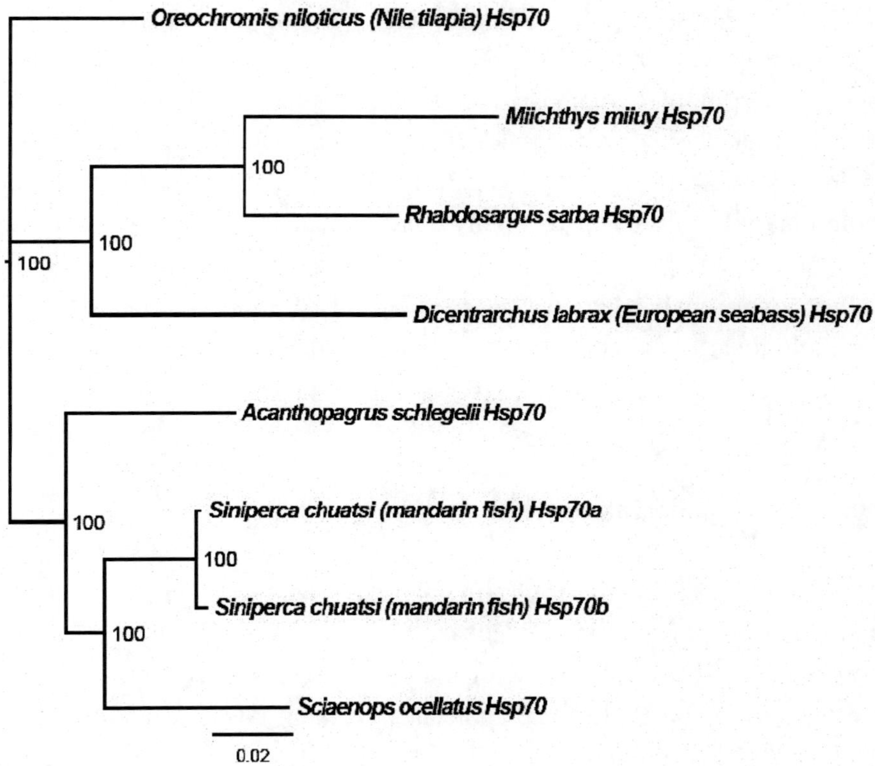

Figure 9. Phylogenetic relationships among Hsp70 amino acid sequences using BI method (arithmetic mean =-3535.337; harmonic mean =-3544.737). Posterior probability values higher than 50% are indicated on each node. The scale for branch length (0.02 substitution/site) is shown below the tree.

However, the outgroup sequence was separated by the other sequences in each of the phylogenetic tree, which were grouped in three and two clusters in the coding cDNA based and amino acid sequences based phylogenetic tree, respectively. These phylogenetic analyses results with the taxonomic data of NCBI Taxonomy database (www.ncbi.nlm.nih.gov/Taxonomy/). In the amino acid sequences based phylogenetic tree the Hsp70 sequences of Miiuy Croaker (*Miichthys miiuy*), gold lined seabream (*Rhabdosargus sarba*) and European bass (*Dicentrarchus labrax*) were grouped together, but only *M.miiuy* and *D. labrax* are members of *Percomopharia incertae*, *R. sarbai*s a member of Spariformes. The other cluster was composed by Hsp70 sequences of blackhead seabream

(*Acanthopagrus schlegelii*), Chinese perch (*Siniperca chuatsi*) and eyed hawk-moth (*Smerinthus ocellatus*). Only *A. schlegelii* is a member of Spariformes. In these comparisons emerged many discordances, which indicated that these organisms Hsp70 phylogeny didn't respect organism phylogeny. Kominek and colleagues data [316] revealed that the evolutionary history of this is highly conserved and this ubiquitous protein family is surprising complex and dynamic. For this reason, further genomic and proteomic analyses are needed in order to have more information about the evolutionary history of these particular proteins in the analyzed fish species

3.6.4. Superoxide Dismutases

Superoxide dismutases (SODs; EC 1.15.1.1) are metalloenzymes widely distributed in prokaryotic and eukaryotic cells [317]. They constitute a protein family that catalyzes the dismutation of superoxide anion ($\bullet O_2^-$) to hydrogen peroxide (H_2O_2) and molecular oxygen (McCord andFridovich 1988). The main physiological role of these proteins is believed to be the protection of the cells by scavenging ($\bullet O_2^-$), which are produced under oxidative conditions by energy generating processes [318, 319]. Depending on species, there may be up to three different metal-containing SOD enzymes in the cells, acting in mitochondrion, nucleus, cytoplasm and extracellular spaces.

These SODs are the products of different genes and they are classified according to the redox metal in the active site. Copper/zinc-containing SODs (Cu,Zn SODs) use copper for the catalytic reaction and they are found in the cytoplasm of eukaryotes, in the chloroplasts of some plants and in the periplasmic space of many bacteria. Furthermore, in eukaryotes, an extracellular SOD (EC-SOD) distinct from cytosolic Cu, Zn SOD has been found [320]. Iron- and manganese-containing SODs (Fe SOD and Mn SOD) bind specifically Fe or Mn [321, 322]. Fe SODs are found in prokaryotes, plants (chloroplasts), and protozoa, while Mn SODs are present both in prokaryotes and mitochondrial matrix of eukaryotes. Numerous fungi possess both cytosolic and mitochondrial MnSODs [323]. A few classes of bacteria are also known to possess activity with either iron or manganese incorporated in the same protein moiety (cambialistic SODs) [324, 325]. Mn SOD and Fe SOD appear as homodimers or homotetramers and exhibit a high degree of sequence and structural similarity [326, 327, 328]; it strongly suggested that these enzymes originate from a common ancestry. In contrast, Cu, Zn and Mn/Fe SODs never shared a common ancestry, and this is shown by their distinctive amino acid sequences [329] and completely different tertiary structures [330]. The discovery of a Ni SOD in Streptomyces and cyanobacteria [331] established a completely new SOD group with a unique Ni-hook motif.

Copper is an essential trace element that plays a vital role as a catalytic co-factor fora variety of metalloenzymes. Beside superoxide dismutase, other enzymes are cytochromec oxidase (mitochondrial electron transport chain), tyrosinase (pigmentation), peptidylglycine alpha amidating monooxygenase (PAM) (neuropeptide and peptide hormone processing) and lysyl oxidase (collagen maturation) [332, 333]. Moreover, copper is redox active metal that can participate in electron transfer reactions, with the consequent production of oxidants, which are able to oxidize cell components [334]. Copper catalyzes the formation of the highly reactive hydroxyl radicals from hydrogen peroxide (H_2O_2) via the Haber–Weiss reaction and it decomposes lipid peroxides to peroxyl and alkoxyl radicals, which favor the propagation of lipid oxidation [335]. Recently, Ferro and colleagues [336] reported the molecular characterization and gene expression analysis of two out of three Cu, Zn SODs from the ciliated protozoon *Tetrahymena thermophila*. In order to increase the physiological response, the cells have been cultured in presence of an excess of copper, which is used as pro-oxidant. These data [336] suggest that *Tetrahymena tt-sod1a* and *tt-sod1b* are the components of an efficient detoxification pathway in continued, elevated presence of metals in the environment. With the aim of evaluating differences at molecular level in relation to their evolutionary history, Ferro and colleagues [336] characterized Cu,Zn SOD sequences from *Oxytricha trifallax* [337] and *Euplotes focardii*, an Antarctic psychrophilic ciliate [338, 339], and compared them with those from other

Alveolata species. In this context, it is particularly interesting that Cu,Zn SODs appear to have evolved rapidly in relatively recent times, whereas Mn and Fe SODs appear to have evolved at a relatively constant rate over the entire history of eukaryotes, [340].

The presence of three genes that potentially encode cytosolic SOD is unique, in unicellular eukaryotes as well as in metazoans [341]. However, Ferro and colleagues [336] results suggest that the functionality of one of these three genes, the *tt-sod1x*, could be limited, since its expression levels are relatively low, as also reported in the Tetrahymena Functional Genome Database (http://tfgd.ihb.ac.cn/). This characteristic is typical of gene families showing a redundancy due to gene duplication phenomena in their evolutionin eukaryotic genomes, and in mammals particularly [342, 343]. For example, in the human genome thirteen metallothi-onein (MT) genes are present, but only ten of them were shown to be functional [344]. Also in *T. thermophila*, duplication processes have driven the evolution of several proteins, including MTs [345, 346, 347, 348, 348, 349, 350, 351]. Regarding Cu,Zn SODs, two two pseudogenes are known to be present on human chromosomes 8 and 16, in addition to the functional human SOD1 gene that is located on chromosome 21 [352]. Although *tt-sod1x* mRNA levels are low, it is possible that this gene is activated under particular stress conditions. This hypothesis is consistent with the phylogenetic relationships of *T. thermophile* Cu, Zn SODs. Ferro and colleagues [336] analysis indicated that *tt-sod1x* emerges together with the Cu, Zn SODs of *Paramecium tetraurelia*. The genome of this protozoan contains two Cu, Zn SOD genes, one of them showing high similarity to *T. thermophila sod1b*. The presence of two separated clusters of Cu, Zn SOD from closely related species can be explained with an evolutionary process characterized by a dupli-cation event that involved the ancestor gene, which probably occurred in distant time, before the speciation event that led to these two species. From Ferro and colleagues [336] data it is impossible to deduce if SOD1a resulted from another duplication event involving the ancestor of SOD1b or SOD1x. This unclear correlation suggests an alternative hypothesis of adaptive convergence that could lead the occurrence of structurally and functionally similar proteins from different ancestor proteins. If it is the case, this process could have happened many times during speciation. This is probably the case of Mn SOD occurrence. In fact, phylogenetic studies placed Mn SODs far from Cu, ZnSODs [353]. Preliminary results indicated that in ciliated protozoa the evolutionary relationships between SOD1s and SOD2s are similar to those of other eukaryotes.

An interesting result from homology modeling analysis is the similarity between *Tetrahymena* SOD1b and *Solanumly copersicum* chloroplastic SOD1. In plants thereare two different groups of Cu, Zn SODs. The first group consists of cytoplasmic and periplas-mic forms, which are homodimeric. The second group comprises the chloroplastic and extracellular Cu,Zn SODs, which are homotetrameric [354]. These enzymes are all encoded by different genes. Even animal cells export a tetrameric glycosylated Cu,Zn SOD that appearsto be more closely related to fungal Cu,Zn SODs, raising the possibility that the tetrameric version is the more ancient one [355]. All these data suggest an independent evolution for *Tetrahymena* SOD1a and SOD2a, and the latter could have a common ancestor with the tetrameric SODs. However, it would be interesting to also investigate further the *Tetrahymena* SOD1x, from both evolutionary and functional points of view.

In fact, our analysis of homology modeling demonstrated structural similarities of SOD1x with the Cu,Zn SOD from *Bombyx mori*. Recent studies revealed that this silk worm enzyme shares 63% sequence identity with human SOD1, with some different residues at mutation sites in Cu,Zn SOD associated to the amyotrophic lateral sclerosis condition [356]. The two expressed Cu,Zn SOD genes code for two proteins that are bigger than the typical IC-SODs (154 aa).

The acquisition of proteins with long primary sequences is a characteristic of the molecular evolution in *Tetrahymena* species (Boldrin et al. 2006) and could explain the high resistance grade of these organisms to environmental stress conditions. In fact, an abnormally long sequence characterizes the Cu,Zn SODs of animals living in permanent stress conditions, such as the icefish. These organisms evolved in oceanic waters surrounding the Antarctica, exposed to high oxygen concentration as a consequence of low temperature, which increases gas solubility.

Furthermore, the inability of icefish to synthesize hemoglobin implies an increased quantity of oxygen simply dissolved in their tissues. These two conditions increase the probability of ROS formation.

In the case of icefish the evolutionary mechanism resulting in long proteins seems to be the mutation of stop codon (TAA → ATG). In *T. thermophila* and *E. focardii* the mechanism probably also involved the stop codons, but this is a peculiar feature of the modified genetic codes of ciliates, where TAA and TAG are not stop codons, but code for *Gln* in *Tetrahymena*, and TGA encodes *Cys* in *Euplotes*.

Despite the different protein dimensions, the residues important for catalytic activity are highly conserved. One exception is *His154* in the isoform *b*, that is replaced by *Asn* through a point mutation that involved the first codon nucleotide (c → a). However, the copper coordination could be maintained by *His165*, which could vicariate the absent amino acid. It is of note that this histidine is also present, at the same position, in *E. focardii SOD1b* (Fig. 2). The only other exception, also in SOD1b, is the mutation of *Thr171*, which is close to the active site channel and, together with the positively charged guanidinium group of Arg177 (strictly conserved), guides the superoxide anion toward the catalytic Cu^{2+} coordination site, stabilizing its bonding. In this isoform it is substituted by Asn by a point mutation, that involved the second codon nucleotide (c → a). In this case, *Thr170* could vicariate the absent amino acid. The other amino acids are fully conserved in both enzymes, such as the coordination sphere of the zinc ion that is formed by three histidine and one aspartic acid residues, assuring the enzyme functionality despite its lower level of sequence identity with other Cu,Zn SOD sequences.

An interesting result is the different pattern of mRNA accumulation observed between *tt-sod1a* and *tt-sod1b*. In particular, the latter appears to be characterized by increasing mRNA levels during the exponential phase of growth, this parameter decreasing after prolonged starvation (24 and 48 h time points). This result suggests that, although the cells are not exposed to an environmental stress factor, an endogenous stress condition could be present. During the exponential growth of the culture, the cells are in active metabolism and the rate of oxygen consumption is high. In this condition,

ROS formation can increase [357] leading to the induction of genes involved in the antioxidant defense, such as superoxide dismutases, but also catalase and glutathione peroxidase, enzymes that are functionally related to SODs, because they eliminate H_2O_2. Between the two Cu, Zn SODs, isoform b seems to be the one that, during evolution, assumed major importance in countering this kind of stress. The exposure of *Tetrahymena* to a prooxidant, such as copper, only slightly increases Cu,Zn SODmRNA levels. This result can be related to the absence of regulatory sequences, such as Metal Responsive Elements (MRE) and ARE, in the promoter regions, or to the low copper accumulation in the cells, or to the low ROS formation rate during the metal exposure. It is well known that the presence of MREs and AREs in the promoter regions is a characteristic of both eukaryotic SOD as well as MT genes [358, 359]. Nevertheless, it has been shown that *Tetrahymena* MT gene transcription induced by cadmium is driven by promoter sequences, which are different from the typical MREs [345, 346, 348]. Therefore, it is possible that Cu-dependent induction is present, but specific metal response elements in Tetrahymena have not yet been identified. Instead, in the promoter regions are present several putative h-ARE sequences, which could probably mediate the induction of SODs by ROS.

After Cu-treatment, the cytoplasmic concentration of this metal significantly increased until 24 h of exposure, afterwards the Cu levels slightly decreased, probably as the result of metal detoxification process, which is dependent on the biosynthesis of glutathione and MTs. Previous studies demonstrate a linear correlation between copper and MT concentrations in the cell-free extract of *Tetrahymena* experimentally exposed to this metal ion [360, 361, 349]. This homeostatic response can limit the presence of copper in the cells, which is not completely removed, but it remains higher than controls even after 48 h. Therefore, copper continuously enters in the cells, but it is not bioavailable to interact with gene regulation systems or to catalyze the production of ROS, because almost all Cu ions are bound by MTs and GSH and continuously eliminated via lysosome system. This could justify the poor induction of SOD gene activation. Nevertheless, SOD activity in Cu-treated cells was found to be increased in comparison to controls, and this was especially evident in the first phases of exposure (0.5-1 h). The absence of a correlation between mRNA levels and proteins activity is difficult to interpret, but it is not a rare phenomenon.

Ferro and colleagues [336] formulated two hypotheses to explain the increase of SOD activity without increase of mRNA concentrations: i) the presence of high Cu-concentrations at cytoplasmic level can be conveyed to SOD apo proteins already present in the cell; ii) copper increases the translation of Cu,Zn SOD mRNA already present in the cell. The mRNAs have a complex lifecycle and non-translating mRNAs, under stress conditions, can be stored in cytoplasmic granules [362], processing P-bodies, and stress granules for decay or future translation, respectively [363]. The first hypothesis could be more plausible, because copper is a fundamental component of the Cu, Zn SOD-active site and it has been supposed that post-transcriptional regulation of Cu, Zn SOD induction is an important mechanism for oxidative stress tolerance in plants [364]. However, it cannot exclude that the little, but statistically significant increase in *tt-sod1a* and *tt-sod1b* mRNA accumulation measured at 1 h, is related also to an increase of ROS production. This hypothesis is supported by preliminary results obtained by ROS quantification experiments in *T. thermophila*, where an intense fluorescence can be observed after copper exposure, probably related to an increase of $\bullet O_2^-$formation induced by this metal.

So far, protozoa seemed to lack the Cu, Zn SOD, a unique feature that would distinguish them from plants, animals and fungi. The hypothesis was that the ancestor of protozoa could have lost these genes during evolution [341]. In Ferro et al 2015 work, it was demonstrated that transcriptionally active genes encoding Cu, Zn SODs are also in protozoa and that the process of gene expression is active. To better explain the physiological role of Cu,Zn SODs in the antioxidant defense system of Tetrahymena, studies on genes coding for other antioxidant enzymes will be necessary, with the aim to reconstruct at both physiological and evolutionary levels, the complex metabolic pathways involved in the cell homeostasis of ROS.

Furthermore, the production of ROS is a prominent feature of innate immune responses. For example, after oral infection, insect hosts can heavily depend on an increase of ROS in the gut lumen, often in concert with a local up-regulation of antimicrobial peptides (AMPs) [365, 366, 367, 368]. Moreover, during bacterial antigen recognition by phagocytes, both ROS and nitrogen oxide are produced as a consequence of phagocytosis, in order to act as cytotoxins and second messengers [369].

In vertebrate acquired immunity, ROS take an active role in the determination of T-cell states [370] and moreover, T cells are also affected by changes in the regulation of the antioxidant system within the phagocytic cells, especially non-enzymatic ROS scavengers such as GSH [371, 372]. Additionally, enzymatic antioxidant system components, such as SOD enzymes, are up-regulated during T-cell death and activation, as a response to increased mitochondrial superoxide ($\bullet O_2^-$) formation [373]. In summary, the role of ROS for infection dynamics in general, and host immune modulation in particular, is essential.

Although innate immunity is a feature of virtually all known multicellular organisms, acquired immunity (i.e., plastic immune responses based on specific memory) is mainly considered a hallmark of vertebrate species. However, recent studies demonstrated that invertebrates are also capable of mounting alternative forms of adaptive immune responses despite lacking features of classical vertebrate immunity [374, 375].

Amongst others, one prominent example is the red flour beetle (*Tribolium castaneum*). In this model organism, enhanced protection after repeated exposure to the same bacterial pathogen was demonstrated for septic as well as oral infection routes [376, 377], and could even be transferred to offspring [378, 379]. This phenomenon is referred to as immune priming, and it has now been reported in a wide range of invertebrate species [374], including copepods, water fleas, cockroaches, snails, shrimps, and mealworm beetles [375, 380]. Although immune priming in these species is well studied, the molecular machinery underpinning immune priming is not well understood. Only recently the identification and characterisation of immune priming-related pathways has begun [374, 381, 382, 383]. For example, the involvement of AMDs, as well as phagocytosis, in immune priming has been experimentally described in *Drosophila melanogaster*, *Tenebrio molitor*, and *Galleria mellonella* [384, 385, 386, 387].

In *T. castaneum*, the analysis of transcriptomic data from two populations of *T. castaneum*, infected with the entomopathogen *Bacillus thuringiensis* by septic wounding, emphasised the importance of an up-regulation of transcription in genes with a predicted oxidative function for the systemic immune response [366]. Another recent transcriptomic study in the same system, but using oral priming by exposure to bacteria-free spore culture supernatants showed that ROS genes were already up-regulated after priming, even when not followed by infection with live spores [388]. This finding encouraged Ferro and colleagues [389] to explore the ROS metabolism of *T. castaneum* during septic priming with this micro-parasite, and furthermore the potential role of ROS-detoxifying genes in invertebrate immune memory.

Ferro and colleagues [389] work was focussed on the characterisation of Cu, Zn-SOD genes combining bioinformatics, biomolecular, and biochemical techniques, in order to gain first data on the evolutionary history of Cu,Zn SODs in *T. castaneum.* Moreover, for the first time, they characterised the involvement of Cu, Zn SODs in physiological responses to immune priming, studying gene expression at both transcriptional and translational levels, in a full-factorial priming and challenge experiment using *B. thuringiensis*.

The deduced amino acid sequences of the four identified Cu, Zn SODs from *T. castaneum* were split into two groups for analyses of the predicted protein structure: *tc-soda* was found to correspond to other metazoan IC-SODs, whereas *tc-sodb1*, *tc-sodb2*, and *tc- sodc* were found belonging to metazoan EC-SODs. Accordingly, Ferro and colleagues [389] identified a putative cleavage site for EC-SOD between residues 16 and 17 (TLA-TV) in Tc-SODb1 and 2, as well as between residues 19 and 20 (VQS-AS) in Tc-SODc, but not in Tc-SODa. Following the initial analysis, the predicted protein sequences of the considered SOD isoforms were aligned with the respective orthologues from other metazoan species with a confirmed crystallography structure. Tc-SODa was compared with SOD1 sequences from *Xenopus lae-vis* (NCBI NP_001080933.1), *Homo sapiens* (NCBI ABL96616.1), and *Bos taurus* (PDB 1CB4), while Tc-SODb1, Tc-SODb2, and Tc-SODc were compared with SOD3 from *Homo sapiens* only, as this is the only metazoan species with a confirmed crystallographic structure for the EC-SOD (PDB 2JLP). The copper and zinc coordination environment and the active site are conserved in all compared sequences for SOD1 and 3. The 3/10 α-helix in the *C*-terminus is missing from Tc-SODb2 and Tc-SODc.

The results of these scientific work [389] showed that the expression of Cu, Zn-SOD genes is affected by immune priming and challenge in *T. castaneum*. Moreover, their findings support the current idea that priming is a specific reaction, that is qualitatively different from a simple low-level immune activation [383, 388] and even involves the overall physiological response. The maintenance of ROS homoeostasis in eukaryotes, is an evolutionary ancient mechanism [390], that is, coadapted as a vital part of innate immunity. This might be another indication that the production of ROS is not only harmful, but can also be desirable to the host [391], especially in an infection scenario. From the recognised potential involvement of the antioxidant system in immune priming, many new questions arise that call for future investigations, e.g., the potential role of peroxide- induced signalling. Indeed, further functional characterisation of the various identified SOD isoforms (as well as other components of the antioxidant system), for example, by RNAi could elucidate their role for immune priming.

3.6.5. Glutathione Peroxidases

Glutathione peroxidases (GPxs; EC 1.11.1.9 and EC 1.11.1.12) constitute a protein family that catalyzes the reduction of H_2O_2 or organic hydroperoxides to water or corresponding alcohols, using reduced glutathione (GSH) as an electron donor [392, 393, 394]. This family comprises more than hundreds members spread over all domains of life [395]. In vertebrates, up to 8 distinct GPxs were identified. Most of them are selenoproteins (mammalian GPx-1, GPx-2, GPx-3, GPx-4, and human GPx-6), while, in the remaining isoforms, the selenocysteine residue (Sec) is replaced by cysteine. Selenium-containing GPxs (Se-GPxs) were also discovered in other organisms such as parasitic trematodes, protists, and bacteria [396, 397]. However, the overwhelming majority of non-vertebrate GPxs contain cysteine in their active site [398]. Previous studies suggest that the evolutionary history of GPx genes is highly complex, because it is the result of a number of independent evolutionary events. They may also be associated to functional convergence processes [399]. Phylogenetic analyses revealed that, in mammals, the GPx gene family evolved from a common ancestral gene by duplication events, followed by random mobilization in the genome. It has also been postulated that GPx-1, -2, -3, -5, and -6 may have a common ancestral gene, probably distinct from GPx-4 [399]. Recently, Sattin and colleagues [400] reported the molecular characterization and gene expression analysis of a new GPx-1 from the Antarctic fish *Trematomus bernacchii*, which represents one of the most studies fish species inhabiting the Antarctic sea, where fish species are exposed to year round low temperatures [401]: on annual basis, the temperature slightly changes around the average value of -1.18 ± 0.68C [402]. In this condition, the dissolved oxygen concentration greatly increases in Antarctic waters, producing a consequent increase of pO_2 in tissues and cells. Therefore, the rate of formation of products from the partial reduction of oxygen and reactive oxygen species (ROS, i.e., O_2^-, H_2O_2 and OH) can increase as the oxygen concentration increases [403]. ROS are physiologically produced in a series of biochemical reactions within cellular compartments (mostly mitochondria and endoplasmic reticulum), but their increased levels may lead to irreversible cell damage and eventually to cell death. Both enzymes and non-enzymatic compounds of the antioxidant systems are present in cells in order to prevent the damages caused by ROS. Like it was previously mentioned, the antioxidant enzymes such as superoxide dismutase (SOD), catalase (CAT), glutathione peroxidase (GPx), and peroxiredoxin (Prdx) act together limiting the effects of oxidant molecules on cells and tissues and constitute the first line of defense. To expand the knowledge on the evolution and adaptation of the GPx-1 s within the group of Antarctic fish, Sattin and colleagues [400] presented the cDNA sequences of this enzyme in other three species, belonging to two families—Nototheniidae (*Trematomus eulepidotus*, *Trematomus lepidorhinus*) and Bathydraconidae (*Cygnodraco mawsoni*).

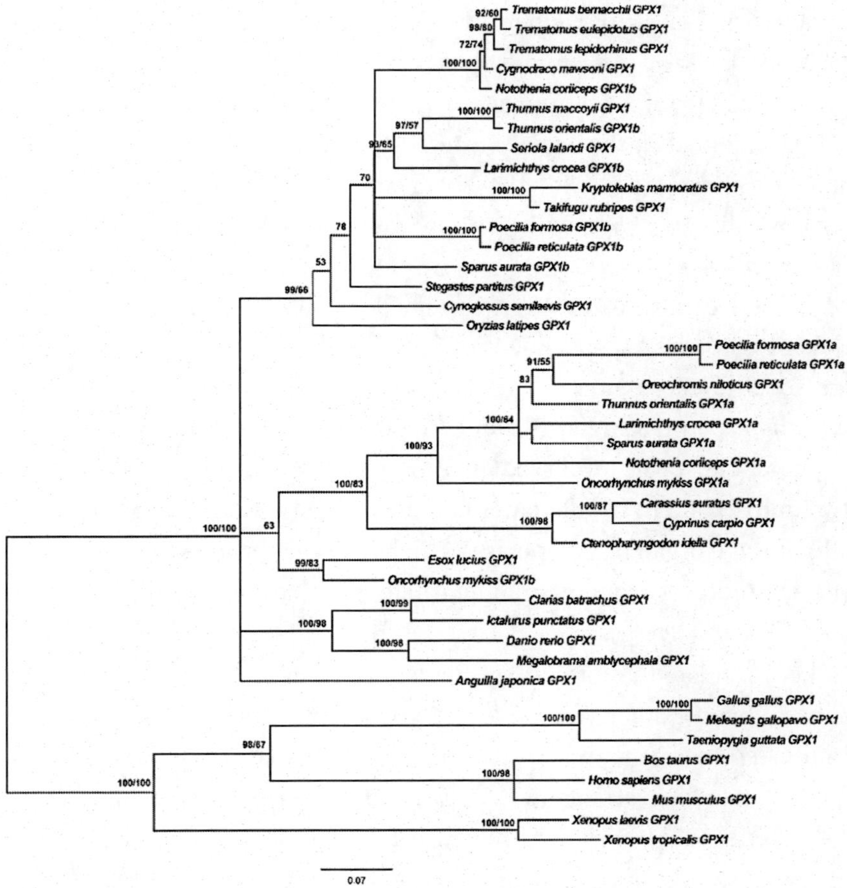

Figure 10. Phylogenetic relationships among GPx1s of various organisms reconstructed on the basis of cDNA sequences and using both BI (arithmetic mean = -10048.93; harmonic mean = -10082.81) and ML (arithmetic mean = -10002.5) methods. Posterior probability (first number) and bootstrap values (second number, if present) higher than 50% are indicated on each node. The scale for branch length (0.07 substitution/site) is shown below the tree.

Bayesian Inference (BI) and Maximum Likelihood (ML) methods generated phylogenies with the same topology, depicted in the cladogram of Figure 10. All fish GPx-1 s are grouped together and clearly separated from other vertebrate GPx-1 s (posterior probability = 100%; bootstrap values = 100%). Among these sequences, the Antarctic fish GPx-1 s obtained in the present work emerge together in a clearly separated cluster (posterior probability = 100%; bootstrap values = 100%). In particular, they are more closely related to GPx-1b from another Antarctic fish, *Notothenia coriiceps*, which is included in the same cluster. All of these sequences cluster with almost all known fish GPx-1b sequences (posterior probability = 96%; bootstrap values = 89%), with the only exception of GPx-1b from rainbow trout (*Oncohynchus mykiss*).

Basal gene expression was determined in gills, heart, liver, and skeletal muscle of non-stressed specimens of *T. bernacchii*, to verify the presence of tissue-specific expression of GPx-1. Figure 11 shows that liver and heart tissues displayed the highest level of mRNA accumulation, approximately four times the levels observed in gills and skeletal muscle ($P < 0.05$). Between liver and heart, the latter had the highest mRNA levels ($P < 0.05$). The liver is the organ that displays the highest level of active Se-GPx protein, about one order of magnitude higher than other considered organs (Figure 12).

The considered Antarctic species appeared to encode a single GPx-1 variant, like most other vertebrate species. Despite only one GPx-1 isoform for most fish species represented in the NCBI database, more recent studies had annotated two variants in several different teleost species, and also in the Antarctic *N. coriiceps*. Although the possibility existed that a second variant had simply not yet been identified in tetrapods, given the growing number of genome sequences available for these vertebrates that seemed unlikely. These data suggested the appearance of the two *gpx-1* genes through a duplication event, likely occurred after the origin of tetrapods, but sometime before the speciation that led to the differentiation of fish families [404, 405, 406]. Based on the results of phylogenetic analyses, Sattin and colleagues [400] created the hypothesis that a duplication event migh have occurred early in the fish evolutionary history. In fact, many species belonging to different orders (*Cyprinodontiformes, Spariformes, Perciformes* and *Salmoniformes*) have two GPx-1 variants. The evolution of these variants seems to proceed together with the evolution of fish orders and families.

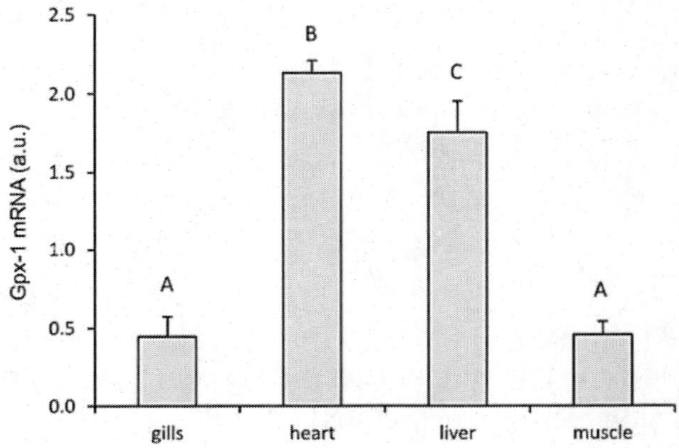

Figure 11. GPx-1 mRNA levels in gills, heart, liver, and skeletal muscle of *Trematomus bernacchii*. Letters: Student–Newman–Keuls test with respect to tissues (P < 0.05).

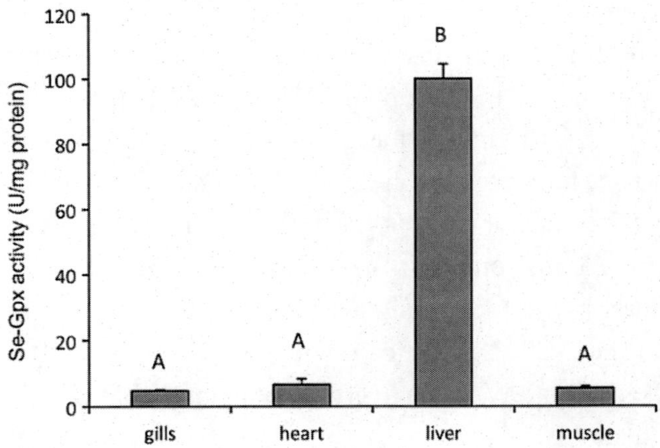

Figure 12. Se-GPx activity levels in gills, heart, liver, and skeletal muscle of *Trematomus bernacchii*. Letters: Student–Newman– Keuls test with respect to tissues (P < 0.05).

Furthermore, in the phylogenetic analysis of coding sequences, the topology of the GPx-1a branch is very similar to that of GPx-1b, but with one important exception represented by *O. mykiss* GPx-1b. This sequence appears more phylogenetically related to GPx-1as than GPx-1bs. This result may suggest the appearance of gpx-1b gene at different times during the GPx-1 evolution, through independent duplication events. However, from the data available so far, it is not possible to safely discuss whether the hypothetical independent duplication events of *gpx-1* could be a distinguishing feature of *Salmoniformes* order or might indeed have happened also in other fish orders. In this respect, it is interesting to note that an extra round of whole genome duplication (WGD) may have occurred in salmonids [407]. The comparison of GPx-1b variant could be related to this event, although the researchers share the view that the WGD is probably responsible for the multiplicity of isoforms (Gpx-1, Gpx-4, Gpx-7) rather than for the multiplicity of variants [408].

The GPx-1s of the Antarctic fish emerge together and result clearly separated from other orthologs and paralogs. The phylogenetic distance between the GPx1 s of *Cygnodraco mawsoni* and *Trematomus* representatives is very small, probably related to the relatively recent divergence of these organisms (11.6–5.3 million years ago; [409]), if compared to the inferred time of appearance of extant fish (beginning around 490 million years ago). In fact, the Antarctic continent was subjected to a series of tectonic and oceanographic events correlated with the formation of the Antarctic Circumpolar Current (ACC) [410, 411]. This oceanic current, established 22–25 million years ago, surrounds the isolated island and creates a natural barrier delimiting a unique evolutionary site between the continent and the ACC. Antarctica gradually became isolated and colder, and new niches became available to groups of animals that were diversifying *in situ* (notothenioids) or immigrating into the Southern ocean (liparids and zoarcids) [412].

Although several studies have focused on the analysis of GPx activity in the Antarctic fish, the gene expression results presented in the Sattin and colleagues [400] work are the first that consider the mRNA accumulation. The gene expression analyses in *T. bernacchii* highlighted high variability in tissue GPx-1 mRNA accumulation, with heart and liver showing the highest transcription levels. This result is consistent with previous studies [413, 414, 415] in which high levels of active Se-GPx protein have been measured in the same organs of other Antarctic Nototheniidae species (*Notothenia rossii*, *Dissostichus eleginoides*, and *Trematomus newnesi*). Most of the Antarctic fish possess substantial body lipid accumulations, considered to be involved in buoyancy and aerobic metabolism [416, 417]. This accumulation has led to a prevailing view that lipid is the primary fuel for energy metabolism of these species. This idea has received direct experimental support from Sidell [418] and Crockett and Sidell [419], who have shown that aerobic energy metabolism is predominantly dependent upon the oxidation of unsaturated lipids in heart and liver of nototheniids. The fact that unsaturated fatty acids are susceptible to oxygen radical attack, which can produce lipid peroxidation through a chain of oxidative reactions [420], could be related to the relatively high levels of Se-GPx activity of heart and liver compared to gills and muscle in all the studied fish.

Trematomus bernacchii also expresses the highest GPx activity in liver, and this observation confirms previous analyses on the antioxidant defense system of this species [414, 421, 415]. The high liver enzymatic activity is probably related to the intense metabolism of this organ, which eventually leads to the formation of organic and inorganic peroxides, as well as other ROS. Also the relatively high Cd concentrations that characterize the coastal water of Antarctica may be responsible of this condition. In fact, Benedetti and colleagues [422] demonstrated that in Cd-exposed *T. bernacchii* specimens there is an increase of GPx activity (approximately 3-fold higher than in control fish). Despite the high levels of messenger RNA detected in the heart, the GPx activity in this organ is an order of magnitude lower than the hepatic tissue. The absence of a correlation between mRNA levels and proteins activity is not a rare phenomenon. In fact, mRNAs have a complex lifecycle and non-translating mRNAs, in organs where the risk of oxidative stress is high, can be stored in cytoplasmic granules [423], processing P-bodies, and stress granules for decay or future translation, respectively [424]. The high ROS production in heart than in liver could also be related to specific physiological and behavioral characteristics of this species. In fact,

T. bernacchii is a relatively inactive species living on or near the substrate [425], but in some circumstances it increases its locomotor activity, and consequently the cardiac activity. An increased heart activity consumes more oxygen and thus increases the production of ROS. This stress condition leads to an increase of antioxidant gene transcription, but also the translation of stored transcripts, resulting in a more rapid physiological response.

3.6.6. Peroxiredoxins

An unusual antioxidant protein, now called perox-iredoxin (Prx, EC 1.11.1.15), was initially identified on the basis of its capacity to protect proteins from oxidative damage caused by reactive oxygen species (ROS), catalysing the reduction of H_2O_2, to water and alcohol in the presence of dithiotreitol (DTT). Analysis of Prx purified from yeast revealed that it did not contain conventional redox centres such as metals, heme, flavin, or selenocysteine, being very different with respect to other known antioxidant [426, 427]. Later, it has been found that (i) Prxs are present in all bio-logical kingdoms, from bacteria to animals; (ii) two cysteine residues, corresponding to Cys47 and Cys170of yeast Prx, are highly conserved among Prx fam-ily members; (iii) Prxs are homodimers arranged in a head-to-tail orientation; and (iv) Cys47–SH of Prx is specifically oxidised by H_2O_2 to cysteine sulfenic acid (Cys–SOH), which is resolved by reaction with Cys170–SH of the adjacent monomer and it results in the formation of a disulfide link, Cys47–S-S–Cys170 [427]. The conserved Cys residue corresponding to Cys47 of yeast Prx was later referred to as the peroxidatic Cys (CP) to reflect its sensitivity to oxidation by peroxides, and the conserved Cys residue corresponding to Cys170 was designated the resolving Cys (CR) [428]. On the basis of the presence or absence of the CR residue, Prxs are classified into typical 2-Cys, atypical 2-Cys, and 1-Cys Prx subfamilies [429]. Animal cells express six isoforms of Prxs: isoforms from 1 to 4 belong to the typical 2-Cys Prx group; Prx5 belongs to the class of 2-Cys enzymes. Isoform 6 is the only one belonging to 1-Cys Prxs [430]. Prx1 is mainly localised in the cytosol, nucleus and peroxisomes, but it has been found also in serum. Prx2 is present in cytosol and nucleus and it has been shown to bond the cell membrane [431]. Prx3 is located exclusively in the mitochondria [432]. Prx4 has been found in both cytosol and endoplasmic reticulum and contains a leader peptide that is believed to be essential for protein secretion [433]. Prx5 is localised in cytosol, mitochondria and peroxisomes [432]. Prx6 is located in cytosol, vescicles and lysosomes (soRo Kina et al. 2011). Prx localisation is multivarious, being dependent on the cell type, but also on the environmental conditions [434]. Many studies suggest that Prxs are more than just simple antioxidant enzymes. For example Prx6 also acts as phospholipase A2 [435]. Other evidences suggest that Prx oxidation also allows them to function as molecular chaperones [436] and regulates the cell cycle [437]. A very interesting proposed role is encompassed by the floodgate hypothesis [428], in which active Prxs normally keep H_2O_2 low (i.e., a closed floodgate), but under signalling conditions, it causes loss of function via overoxidation in a localised region of the cell, allowing H_2O_2 to build up locally (i.e., be released by an open floodgate) for signalling purposes [438, 439]. The abundance of Prx is high: it can account for up to 1% of all soluble cellular proteins [428]. Furthermore, typical 2-Cys Prxs are the largest and most widely distributed subfamily. These Prxs are moonlighting proteins: at high H_2O_2 concentrations they act as holdases, whereas when the rate of ROS formation is low they are thioredoxin-dependent peroxidases [436]. The dual functions of typical 2-Cys Prxs, modulating ROS concentrations and preventing protein aggregation, may play pivotal roles in cellular response to pathogens and external stress [436]. The importance of Prxs is unarguable, as transgenic/knockout mouse models overexpressing or deficient in most highly expressed Prxs has demonstrated a decrease in genome stability and accelerated aging, and an overexpression of these proteins is associated with various problems related to cancers treatment [440]. Recent studies show that Prxs expressed in tumor cells play positive roles in their progression and/or metastasis in transplanted animals. Different functions of Prxs are required for their progression/metastasis *in vivo* depending on tumor types [441].

Metazoa is one of the great eukaryote kingdoms. An increasingly well resolved Proterozoic fossil record documents the presence of most of the major taxa of eukaryotes, including the rhodophytes, stramenopiles, alveolates and green plants during the late Mesoproterozoic to early Neoproterozoic (about one billion years ago). A coincident rise in a critarch diversity, combined with molecular phylogeny evidence for rapid cladogenesis, points to a major radiation of eukaryote groups at this time, sometimes referred to as the "big bang" of eukaryotic evolution [442]. The most fundamental division within the bilateral metazoans is the protostome-deuterostome branch. Protostomes include at least annelids, arthropods, molluscs and platyhelminths. They have spiral cleavage and usually mosaic development, are schizocelic, form the mouth at (or near) the site of the blastopore and mesoderm from a mesentoblast that is usually 4d. Deuterostomes include at least echinoderms and chordates, the latter group including the backboned animals. They have radial cleavage and usually regulative development, are enterocelic, form the mouth away from the blastopore and mesoderm from endodermal cells along the archeneteron [443]. Approximately 2.5 billion years ago, photosynthetic bacteria acquired the capacity to photodissociate water, leading to the geologically rapid accumulation of molecular oxygen during the Great Oxidation Event (GOE), when anaerobic life underwent a catastrophic decline [444]. Organisms that survived the transition to an aerobic environment were those that respired and/or evolved oxygen. All oxygen-utiliszing organisms acquired constitutive antioxidant defences (both small molecules and enzymes), detoxifying and scavenging the ROS that are continuously produced as a by-product of aerobic metabolism [357, 358]. Among them, Prx has been object of an increasing interest for its pivotal role in cell defense and as conserved marker for circadian rhythms in metabolism. Edgar and colleagues [444] studied the cycles of Prx oxidation-reduction in Eukarya, Bacteria and Archaea and proposed that all these organisms have cellular rhythms sharing a common molecular origin. In fact, it has been proposed that the ability to survive cycles of oxidative stress may have contributed to a selective advantage from the beginnings of aerobic life. Twenty-four hours cycles of Prx oxidation–reduction in all domains were observed in both Archaea,

Bacteria and Eukaryota supporting the hypothesis that cellular rhythms share a common molecular origin [444]. However, only a few studies have been dedicated to the evolution of Prxs [445, 446, 447]. Bakiu and Santovito [448] decided to extend the research on Prx evolution performing phylogenetic analysis on the most studied group of eucariotic organisms (Eukarya), the animals, using the Bayesian approach.

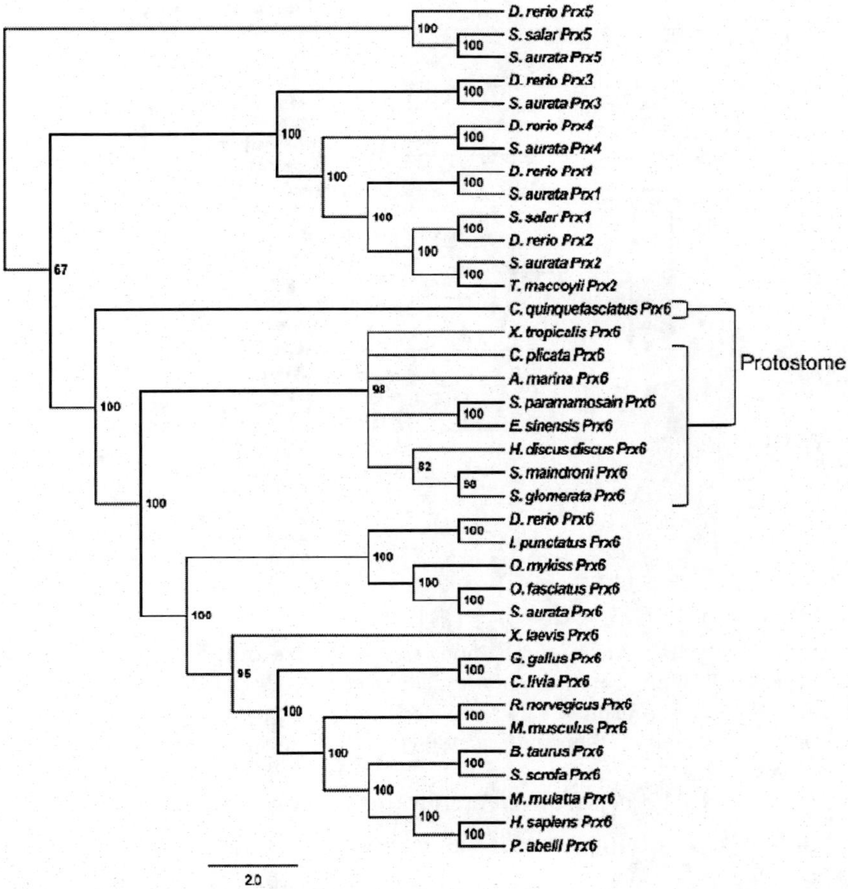

Figure 13. Phylogenetic relationships among metazoan Prxs using coding sequences and BI (Bayesian Inference) method (arithmetic mean = -29057.97; harmonic mean = -29083.98). Posterior probability values higher than 50% are indi-cated on each node. The scale for branch length (2.0 substitution/site) is shown below the tree.

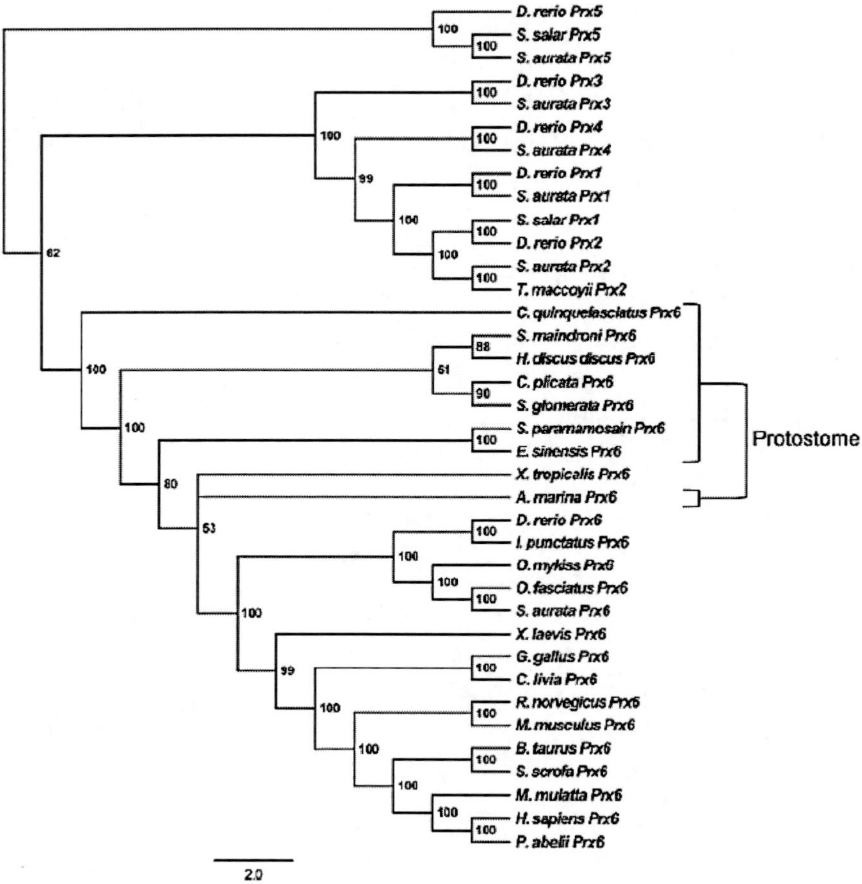

Figure 14. Phylogenetic relationships among metazoan Prxs using amino acid sequences and BI (Bayesian Inference) method (arithmetic mean = -12412.89; harmonic mean = -12437.68). Posterior probability values higher than 50% are indicated on each node. The scale for branch length (2.0 substitution/site) is shown below the tree.

Fortunately, there is sufficient information about many characterised nucleotide and amino acid sequences of metazoan Prxs. Using bioinformatics tools they analysed the phylogenetic relationships among metazoan Prxs, with the aim to acquire new data on the molecular evolution of these proteins. In both cladrograms (Figure 13 and Figure 14), three main clusters were present: one including all typical Prx 2-Cys (isoforms 1, 2, 3 and 4), Prx5 isoforms (atypical 2-Cys Prxs) and the third grouping Prx6 isoforms (1-Cys). This distribution confirmed the previously obtained results using non-Bayesians methods [447]. In particular, the phylogenetic relationships among the typical 2-Cys isoforms were compatible with the hypothesis exposed by Copley and colleagues [445], which suggested an evolutionary model based on the appearance of 3 groups of Prxs (1-2, 3 and 4) through the loss of N-terminal extension in Prx3s and the acquisition of C-terminal extension in Prx4s.

The positive and negative selection sites identified by using a bioinformatic approach (by Bakiu and Santovito) were compared with the crystal structure of Prx1 from *Schistosoma mansoni* (SmPrx1, PDB code 3ztlD, solved by X-ray, resolution 3.00Å) characterised by Saccoccia and colleagues [449]. In the active site of SmPrx1 low-molecular-weight dimer, residues from 47 to 50, including the CP (Cys48), form the first turn of the α2 helix. The compact hydrophobic pocket is formed by Tyr40, Pro41, Ala42, Thr45 and Pro49 that, together with Arg124, surround the CP [449]. Pro49 is highly conserved in typical 2-Cys Prxs [450] and reduces the propensity of the first turn of the α2 helix to retain its secondary structure. Bakiu and Santovito [448] results indicated that only codons 41, 48 and 49 were negatively selected, which confirmed that some (but not all) of the important for peroxidase activity amino acids were conserved during the evolution of Prxs. While these data confirmed the soundness of their results, on the other hand it brought out some questions on the absence of conservation for the other amino acids that were important for catalytic activity. In this subset of sequences from invertebrates, the number of negatively selected codons (about 12% of the 224 codons) was higher than the positively selected ones (approximately 5%).

All these results confirmed that the molecular evolution of metazoan Prx was peculiar and suggested that the positive selection may have operated into the evolution of these proteins and a purifying selection has been present during this process. Probably, the natural selection was responsible for the deviation from the evolution pathway of Prx genes, characteristic of the protostome isoforms. Anyway, further analyses are needed to verify if the positively selected codons, identified by bioinformatic approach in the Bakiu and Santovito [448] study could really codify essential amino acids. A possible verification could be done by site-specific mutagenesis of those nucleotides, that could structurally and functionally affect the enzymatic activity of these proteins. Purifying selection is important for the evolution of a gene family, because it can help the belonging genes to maintain their optimal function.

However, positive selection is an important source of evolutionary innovation and it is a major force underlying the adaptation of species to a new environment [451]. Many proteins have been found under positive selection, such as those involved in immunity (MHC, immunoglobulin VH, class 1 chitinas), proteins or pheromones involved in reproduction (abalone sperm lysin, sea urchin bindin, proteins in mammals) and proteins that acquire new functions after gene duplication [448]. Bakiu and Santovito [448] suggested to include Prxs among those proteins, as their diversification allowed the cells of all organisms to acquire a powerful defence system against the risk of oxidative stress.

Let turn back on some considerations regarding the molecular characterization of peroxiredoxins response to warming. Temperature is a physical parameter that affects many biological processes, especially those related to energy. The main characteristic of Antarctic seawater is its low constant temperature of about −1.8°C that determines a high concentration of dissolved gasses, O_2 in particular, in the environment as well as in animal body fluids. This condition can increase the formation rate of reactive oxygen species (ROS), molecules that are normally produced in cells from the partial reduction of O_2 [358], also increasing the risk of oxidative stress [357]. In the Antarctic fish this stress condition is favored by the presence of body lipid accumulations, considered to be involved in buoyancy and aerobic metabolism of these organisms [416, 417, 420]. Both enzymes and non-enzymatic compounds of the antioxidant systems are present in cells in order to prevent the damages caused by ROS. Like previously mentioned, antioxidant enzymes such as superoxide dismutase (SOD), catalase (CAT), glutathione peroxidase (GPx) and peroxiredoxin (Prdx), act together limiting the effects of oxidant molecules on cells and tis-sues and constitute the first line of defense. Peroxiredoxins (Prdxs, EC 1.11.1.15) are antioxidant enzymes that catalyze the reduction of peroxides using redox-active cysteines [452, 453].

Tolomeo and colleagues [454] did a scientific study, which was focused on the last subfamily, 1-Cys Prdxs, that are represented by peroxiredoxin 6 (Prdx6). It is a bifunctional protein, possessing the activities of both glutathione peroxidase and phospholipase A2 (PLA2) [455, 456]. In particular, this enzyme is characterized by a single peroxidatic cysteine that after oxidation is reduced by glutathione S-transferase-bound GSH to complete the catalytic cycle [457]. The peroxidatic cysteine (Cys47 in mammals) is placed in the catalytic center characterized by the amino acid sequence PVCTTE, analogous to the surrounding sequence in typical 2-Cys Prdxs, FVCPTE [458]. In mammals, the peroxidase activity is due to a catalytic triad, represented by *Thr43* and *Arg130* residues together with the peroxidatic *Cys* introduced above. The PLA2 activity of Prdx6 is blocked by serine protease inhibitors, providing evidence for a serine as the active center. The primary sequence of Prdx6 at amino acids 30–34 indicates a lipase motif (GXSXG), which occurs in essentially all serine-dependent lipases. This motif is not present in other peroxiredoxins. Furthermore, examination of the crystal structure of human Prdx6 suggests that *Ser32* is part of a surface-localized catalytic triad (*His26*, *Ser32*, *Asp140*), that is common for serine-based PLA2. Site-directed mutation of *Ser32* resulted in the loss of PLA2 activity, which was interpreted to confirm a role for this amino acid moiety in catalysis [459]. Prdx6s have received considerable attention in recent years as important cellular antioxidant proteins that help control intracellular peroxide levels. In addition, these proteins may also function, through peroxide-mediated inactivation, in hydrogen peroxide signaling in eukaryotes [439].

Various gene sequences of Prdx6s from aquatic animals have been reported [460, 461, 462], although only few Prdx6 cDNAs have been sequenced from fish [463, 464] and none from Antarctic species. Tolomeo and colleagues (including Bakiu R.) [454] reported on the molecular characterization of two Prdx6s from the emerald rockcod, *Trematomus bernacchii*, a teleost widely distributed in many areas of Antarctica. With the aim of evaluating differences at molecular level in relation to their evolutionary history, they also compared the sequence of Prdx6s from *T. bernacchii* with those from other vertebrate species. In this respect, the evolution of Prdxs is particularly interesting, since they are considered ancient proteins, evolved by duplication events during the past 2.5 billion years together with ROS and metabolic pathways.

In particular, in eukaryotes, the higher number of genes coding for family members is only partly explained by the compartmentalization and the existence of mitochondrial, nuclear, peroxisomal and chloroplast isoforms. Furthermore, Tolomeo and colleagues [454] employed molecular modeling technique to perform a structural reconstruction of the two cold adapted Prdx6s. In particular, using different *in silico* approaches, they investigated Antarctic fish Prdx6s (i.e., compared with other, non-cold-adapted, cytosolic Prdx6s) for the presence of molecular characteristics for cold adaptation. In order to gain some insight into the potential physiological role of the Pdrx6 isoforms in *T. bernacchii*, they characterized basal mRNA levels for both Prdx6A and Prdx6B in various organs and tissues. Furthermore, Tolomeo and colleagues [454] analyzed changes in transcript levels for both *Prdx6* genes after exposer to short-term thermal stress to gain a better understanding of their potential role in the thermal stress response of *T. bernacchii* to increased sea surface temperatures related to global climate change.

Phylogenetic relationships of *Prdx6* cDNA sequences were determined using the most powerful statistical method of BI. BI and ML methods generate phylogenies with the same topology, depicted in the cladogram of Figure 15. *T. bernacchii* Prdx6s emerge in two clearly separated clusters, which include all known fish Prdx6s. In particular, *T. bernacchii* Prdx6A is more closely related to Prdx6s from *Miichthys miiuy* and *Sparus aurata* (posterior probability = 99%; bootstrap values = 58%). *T. bernacchii* Prdx6B is most closely related to a Prdx6 from another Antarctic fish, *Notothenia coriiceps* (posterior probability = 100%; bootstrap values = 100%). These two clusters are clearly separated from Prdx6s from other vertebrates (posterior probability = 96%; bootstrap values = 89%). evolution with a gamma shape value (four rate categories) of 1.0 using all statistical criterion: AIC, cAIC and BIC (−lnL = −7241.91).

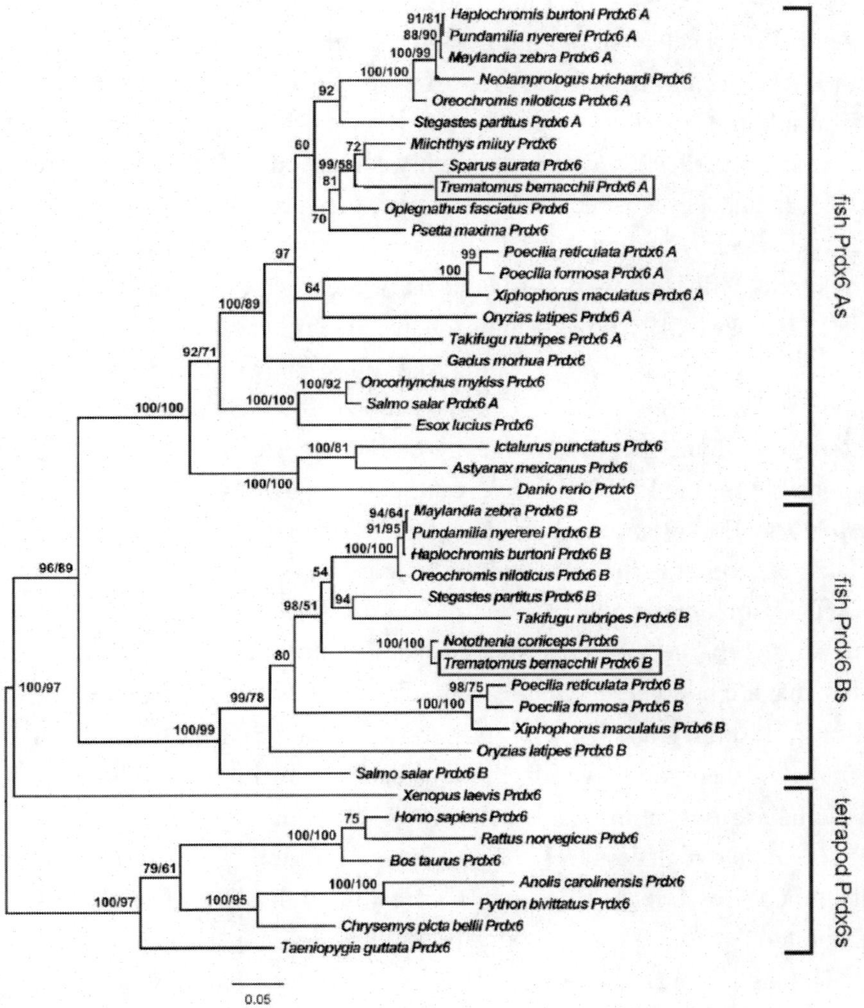

Figure 15. Phylogenetic relationships among different organ-isms Prdx6 reconstructed on the basis of cDNA coding region sequences and using both methods BI (arithmetic mean −11,513.54; harmonic mean −11,549.48) and ML (arithmetic mean −11,353.1). Bayesian posterior probability (first number) and bootstrap values higher than 50% are indicated on each node, respectively. The scale for branch length (0.05 substitution/site) is shown below the tree. *T. bernacchii* Prdx6 are boxed.

MEC analysis indicated a strong prevalence of negative selection in the evolution of Prdx6s, but there are also amino acid residues that demonstrated the effect of positive selection on them. In fact, these analysis highlighted

Glu58, *Arg79*, *Lys92*, *Cys94* and *Cys95* (as referred to in zebrafish Prdx6) to be positively selected. In Figure 16, the deduced amino acid sequences of both *T. bernacchii* Prdx6A and Prdx6B were aligned with Prdx6s from other fish in addition to *Homo sapiens*. Amino acids that are essential for both peroxidase and PLA2 activities are highly conserved in *T. bernacchii* Prdx6s as well as in the other species. In particular, the peroxidatic cysteine (Cys46 in *T. bernacchii*), which is present in the functional pattern PVCTTE and forms a catalytic triad with *Thr43* and an *Arg*, at the position 130 in Prdx6A and 132 in Prdx6B, shows a high degree of conservation across all fish species. The one exception being the only known Prdx6 isoform in *Oncorhynchus mykiss*, where *Glu49* is replaced by Gly in the functional pattern PVCTTE. Similar to the peroxidatic cysteine, the amino acids responsible for the PLA2 activity, represented by *His25*, *Ser31* and *Asp138* (*Asp140* in Prdx6B), are also well conserved (Figure 16). Here, the only notable exception is found within the functional pattern that includes *Ser31* (GXSXG) of *Sparus aurata* Prdx6, where *Gly43* is replaced by *Arg*. In Prdx6A of *T. bernacchii* there are two specific substitutions involving amino acids, that are not included in the catalytic group: *Ile34* is substituted by Val (also in Prdx6B) and *Lys124* is substituted by Ser. In Prdx6B of *T. bernacchii* there are five of these specific substitutions, namely: *Cys* replacing various amino acids in position 74; *Glu92* that is substituted by *Asp*; *Leu* that replaces *Met*, *Ile* or *Val* in position 127; *Leu169* that is substituted by *Ala*; *Ala176* that is substituted by *Ser*; *Pro191* that is substituted by *Ala*.

Though isoform A shows a greater negative isosurface compared to that of isoform B (Figure 17a, b, respectively), the other side of the dimer (Figure 17c, d, respectively) is in favor of a negative surface for isoform B. Isoform B is almost uniformly negative whereas isoform A has a highly negative side (Figure 17a) and a weak positive side (Figure 17c). The latter is also found in *Stegastes partitus* isoforms that have a higher gap of total negative charge between isoform A and B. Template structure, human peroxiredoxin-6, does not have a prevailing negative surface and has a lower otal negative charge compared to other species and positive and negative potentials are equally distributed on its surface.

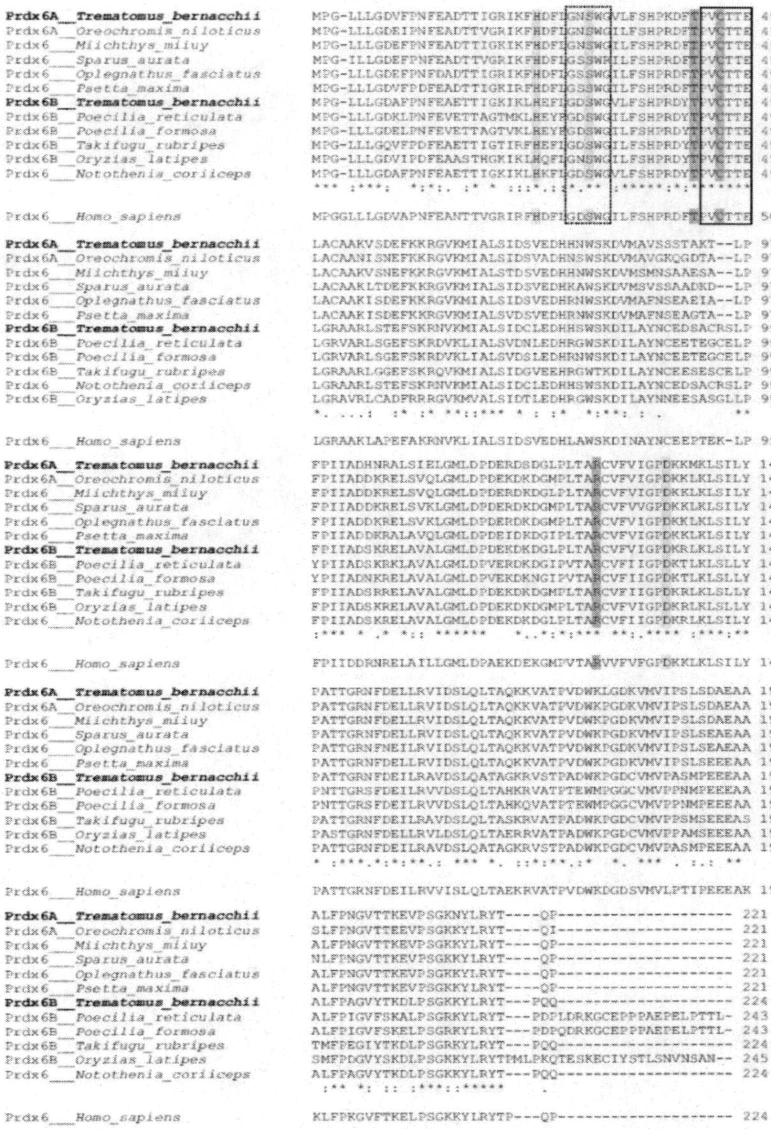

Figure 16. Multiple alignment of the amino acid sequences of 13 Prdx6s from Teleost species as well as from *H. sapiens*. The catalytic centers for peroxidase and phospholipase activities are boxed in continuous and dotted lines, respectively. The amino acid residues highlighted in dark grey represent the peroxidase triad. The amino acids of catalytic triad involved in phospholipase activity is highlighted in light grey. The symbols at the bottom of the teleost sequences correspond to the definitions of the CLUSTAL program: asterisk fully conserved, colon highly conserved, dot conserved substitution.

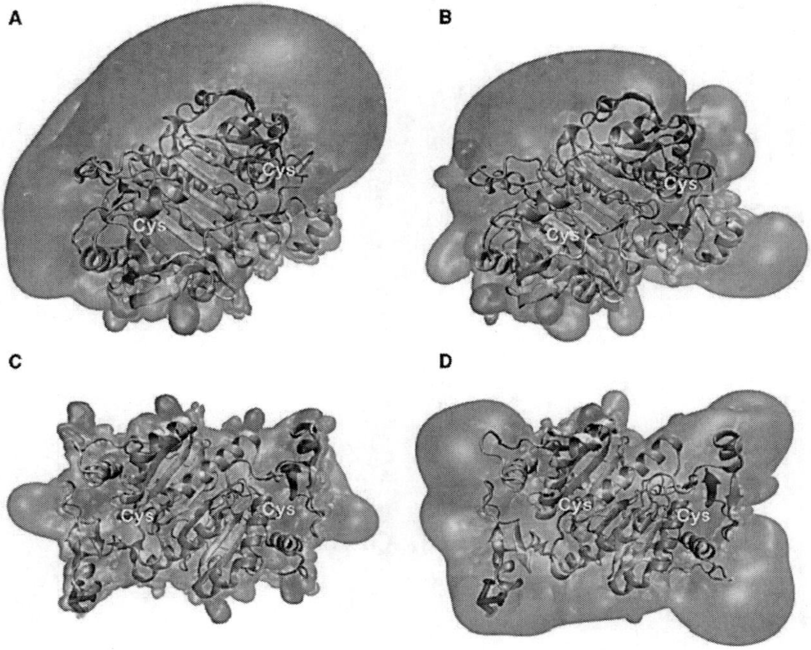

Figure 17. Potential isosurfaces are shown for isoform A (a, c) and isoform B (b, d). Peroxidatic cysteines are reported in ball and sticks.

The presence of two Prdx6 variants in this species seems to be a common feature of fish, unlike most other vertebrate species that appear to encode a single Prdx6 isoform. Although the possibility exists that a second variant has simply not yet been identified in tetrapods, given the growing number of genome sequences available for these vertebrates, this seems unlikely. Despite the fact that only one Prdx6 isoform is described for most fish species represented in the NCBI database, several more recent studies have annotated two variants in several different teleost species. This suggests the appearance of the two *prdx6* genes through a duplication event likely occurred sometime before the speciation that led to the differentiation of fish families (beginning around 490 million years ago).

Furthermore, the evolution of the two variants seems to proceed together with the evolution of fish orders and families. In fact, in the phylogenetic analysis of coding sequences, the topology of the Prdx6A branch is very similar to that of Prdx6B, at least for *Cichliformes* and *Cyprinodontiformes*, in which sequences are grouped in well-supported clusters. Furthermore, Prdx6 sequences of species belonging to *Perciformes*, including *T. bernacchii*, fit this evolutionary pattern, emerging as a sister group of *Cichliformes* Prdx6s for both Prdx6A and Prdx6B. One peculiar exception to these trends are the positioning of *T. rubripes* (order *Tetraodontiformes*) Prdx6s. While variant B emerges with the ortholog from the Perciformes *S. partitus*, variant A emerges with the sequences of other Neoteleostei, yet it shares little correlation with both *Ovalentaria* and *Eupercaria* Prdx6s. An unexpected result came from the comparison between the analyses of nucleotide and amino acid sequences of the Prdx6 isoforms. For Prdx6B, the two topologies were very similar; however, for Prdx6A we noted some important differences. In the *Ovalentaria* taxon the differences are limited, involving only the Prdx6s of *S. partitus*, while in the *Eupercaria* taxon, we found the nucleotide sequences grouped together, but not the amino acid sequences. For instance, *Gadus morua* Prdx6 coding sequence clearly clusters as a sister group of *Ovalentaria* and *Eupercaria* Prdx6s, yet the amino acid analysis revealed weak relationships.

Further analysis suggest that although substitutions at the nucleotide level are limited, a number of them are non-synonymous, leading to the higher variability observed at amino acid level. This could be an indication of a purifying selection effect. However, some residues could be under a strong positive selection, responsible for the higher degree of diversity among the vertebrate Prdx6 amino acid sequences. In particular, the last three residues of the positively selected amino acid sites (*Lys92*, *Cys94* and *Cys95*) are located within the motif FNQDKACC also subject to positive selection. This is a feature that is rarely seen investigating the presence of positive selection, and probably it could easily have contributed to the functional diversification of the two Prdx6 variants in fish. Despite choosing a phylogenetically coherent group of organisms, and a protein that is highly conserved, it was difficult to identify amino acid changes in *T. bernacchii*, that were possibly linked to thermal adaptation. From the data available so far, it is not possible to safely discuss whether the diversification of fish Prdx6s could also be correlated with specific physiological and behavioral characteristics of different families and species. Antarctic fish evolved in this environment, isolated for 10–12 million years, exposed to a very low and constant temperature as well as a high oxygen concentration. It has been proposed that such conditions may have affected their metabolic adaptive strategies during the evolution of this group of teleosts.

Indeed, several unique metabolic adaptations have been observed in these endemic fish, including increased mitochondrial density and the utilization of lipids as a primary energy source [465, 466, 467, 468]. It is thought these adaptations potentially result in an increased susceptibility to reactive oxygen species and oxidative stress [469, 470], and the related evolution of efficient antioxidant systems [471, 472]. The high levels of Prdx6b mRNA measured in heart and liver of *T. bernacchii*, together with the relatively high anti-oxidant enzyme activities measured in the same organs of various Antarctic fish [413, 422, 415], where unsaturated fatty acids are predominantly accumulated, could be an example of this physiological adaptation in not othenioids.

Further support for the perturbing effects of extreme cold temperatures on levels of oxidative damage have recently been published by Place and colleagues, who report a significantly higher level of oxidative damage occurs in multiple tissues of *T. bernacchii* when acclimated to −1.5 °C compared to specimens acclimated to +4 °C [422].

In this perspective, it might be interesting to note that Prxd6A of *T. bernacchii* has a unique structure with a particular surface charge distribution not seen in tetrapod orthologues. Variant A has a highly negative side and a weak positive opposite side, whereas variant B is almost uniformly negative (similar to the surface charge of human Prdx). This situation is also observed in Prdx6 orthologues of *S. partitus*, suggesting this may be a common feature of all fish. Therefore, the presence of two variants of Prdx6s both in the Antarctic fish as well as in those of temperate zones could constitute an early adaptation, that subsequently allowed these poikilothermic aquatic vertebrates to colonize very different environments, characterized by different temperature ranges. In this perspective, the possible thermal specialization of the two Prdx6 variants may have evolved prior the isolation of the Antarctic species, rather than constituting an adaptation to low temperatures, that would have occurred in relatively recent times.

Although we cannot identify specific amino acid substitutions that are directly linked to cold adaptation in the Prdx6 variants, a number of amino acid substitutions unique to *T. bernacchii* suggest Prdx6A may display greater degree of flexibility at cold temperatures. Among these, *Ile34* is substituted by *Val* (this substitution is also present in the Prdx6B of *T. bernacchii* and *N. coriiceps*): both these amino acids are hydrophobic, but *Val* may confer more flexibility than *Ile* [473]. Given the proximity of this substitution to the functional motif that stabilizes *Ser32* (part of the phospholipase catalytic triad) this could potentially affect the phospholipase activity of Prdx6A.

The same argument can be generated for the substitution of *Lys124* by *Ser*, which replaces a charged and rigid one amino acid with a polar, more flexible one [473]. This substitution occurs in a loop region close to *Arg132* (which is part of the peroxidase catalytic triad) and *Asp140* (which is part of the phospholipase catalytic triad) and could therefore affect both enzymatic activi-ties of the protein. Finally, in *T. bernacchii* Prdx6A, *Pro183*, whose ring structure provides greater rigidity, is replaced by *Leu*, which is also a hydrophobic amino acid, but confers greater chain flexibility in another important loop region. Given that single point mutations in loop regions can confer dramatic changes in enzyme kinetics [474], cumulatively, these amino acid substitutions may be related to low temperature adaptation [475, 476].

In addition to biochemical properties, that suggest the Prdx6A orthologue in *T. bernacchii* may display temperature specialization, Tolomeo and colleagues [454] generated gene expression data, which also indicated that Prdx6A seems to be the variant most affected by temperature. When acclimated to temperatures as little as 3 °C above ambient, they observed a significant decrease in the transcript levels for Prdx6A in both liver and heart tissues. Meanwhile, Prdx6B mRNA levels displayed significant increases in these same tissues.

While the down-regulation of Prdx6A could potentially be viewed as an energetic trade-off, this would be counterintuitive to previous studies that have shown *T. bernacchii* experiences increased levels of oxidative damage during short-term thermal stress, likely as a result of elevated metabolic rates [477, 422]. In fact, Tolomeo and colleagues [454] have evidence that transcript levels for cytoplasmic super-oxide dismutase (Cu, ZnSOD) increase in these same fish. Increased Cu, ZnSOD activity, which converts superoxide ions into the less reactive species, H_2O_2, would result in elevated levels of substrate for enzymes such as Prdx6 and selenium glutathione peroxidase (GPx-1), which further reduce H_2O_2. Tolomeo and colleagues [454] also have recent data that demonstrate mRNA levels of GPx-1 also increase in these fish, when combined with the increases in *prdx6b* and *Cu, ZnSOD* transcripts; it suggests that *T. bernacchii* is actively defended against ROS production at a time when Prdx6A is down-regulated.

A more plausible hypothesis is that Prdx6B shares functional overlap with Prdx6A, and, if its structural stability is less temperature sensitive as suggested by the sequence and biochemical analyses, may compensate for the down-regulation of Prdx6A at elevated temperatures. Ideally, it would be best to further test the hypothesis that Prdx6A is a "cold-adapted" enzyme by measuring the thermal stability of the active protein for the two variants of Prdx6s. Unfortunately, at the moment these data are not easy to obtain, since there is currently not an optimized methodology to test the activity of the 1-Cys Prdxs (unlike the 2-Cys Prdxs, for which a commercial kit also exists).

The data of Tolomeo and colleagues [454] are the first on the molecular and functional characterization of the genes encoding the Prdx6s of Antarctic fish, and constitute the starting point to study the expression of these enzymes in specific ecological contexts such as Antarctica. In particular, our expression analyses may be important for predicting climate change responses in this organism. In fact, the obtained results revealed rapid and specific responses of the Antarctic rockcode to temperature. The Tolomeo et al (2015) represented the first demonstration about the effectiveness of using Prdx6s gene expression as biomarkers of stress in environmental field studies related to global warming. Furthermore, the presence of a Prdx (Prdx6B), whose expression is activated by increased temperatures, may be a condition that limits the stenothermy of *T.*

bernacchii, rendering this species less vulnerable to moderate environmental temperature changes and other environmental perturbations associated with global climate change.

3.7. Exploring Gene Regulation and Genomics

Over the past 15 years, numerous studies conducted in aquatic ecotoxicology have widely contributed to our understanding that the most prominent environmental contaminants cause changes in gene expression through activation of receptors or other ligand-activated transcription, particularly for fish models.

In a review considering the basic processes of cell signaling and gene expression explored in *Fundulus heteroclitus*, [478] underlined the major roles played, for example: (1) by AhR, a family of transcription factors regulating the expression of cytochrome P450s and other genes in response to organic pollutants, and (2) by the hypoxia-inducible factor (HIFα) that could be toxicologically relevant.

In the past decade, sequence data have facilitated research in aquatic toxicology. Genomic information can be used to design and develop microarrays for some selected genes to thousands of genes, microarrays being constructed by cDNAs or oligonucleotides, and spotted arrays are commercially available for a few fish species [479].

Microarrays can be used to determine upregulation and downregulation of gene transcripts (mRNA) in exposed organisms. It is widely considered that transcriptomics aids in the identification of new toxicological effects as well as molecular, cellular, and biochemical pathways, but the link between transcript expression changes and effects at higher biological levels has rarely been addressed [479]. Furthermore, Nikinmaa and Rykönen [480] have emphasized the need to relate mRNA and protein levels, and protein activity of individual genes in aquatic toxicology, as studies of mRNA expression do not necessarily reflect protein expression.

The comparison of transcriptome responses between natural populations in the field from polluted and reference sites could be particularly effective to detect contemporary evolution in heavily contaminated systems [481, 181]; in addition to revealing mechanisms of toxicity and tolerance to pollutants, these studies underline the value of comparative transcriptomics to explore adaptive differences among wild populations.

Chapter 4

SENTINEL SPECIES

The increasingly precise techniques of analytical chemistry allow to identify and to quantify, in all biological or physical matrices, an increasing number of xenobiotics and their metabolites, but these measures supply no information about the toxicological or ecotoxicological risks of such contaminants. For that purpose, toxicity tests, carried out in the laboratory on a limited number of species, were established, allowing definition of the nature of the risk as well as any associated dose–effect relationship.

However, these tests do not take into account the influence of the multiple parameters present in the environment, that intervene under natural conditions to affect toxicity: abiotic ecological factors (such as temperature, salinity, conditions of oxygenation), interspecific variability and the interactions between species, the heterogeneity of populations in their interactions with pollutants, or between interactions between pollutants and abiotic factors of the environment [482].

The biological monitoring of the environment, or "biomonitoring," has the objective of integrating these various aspects, particularly by using sentinel species as models.

The notion of sentinel species has existed for a long time: in the nineteenth century, miners took a canary with them down into a mine. The appearance of symptoms of asphyxiation in the bird warned them of the presence in the subterranean galleries of dangerous concentrations of toxic gas. The use and the definition of the term of "sentinel species" evolved over time, and in different contexts, explaining why today there are disagreements and differences in the use of the concept of sentinel species between authors.

Phillips [483] considered that species used for biomonitoring are referred to under the several generic names of "bioindicators," "bioaccumulative indicators," "biomonitors," and "sentinel organisms," and that these names are generally badly used. As often defined today, sentinel species may be considered as any species providing a warning of a dysfunction or an imbalance of the environment, or, more restrictively, a warning of the dangers of substances to human health. In the widest terminology, the sentinel species can be (Figure 18):

- A bioindicator species, providing information by its absence (or presence) and/or of abundance of individuals;
- A bioaccumulative species, the capacity of which to accumulate contaminants at levels much higher than those in the environment, designates it as the material of choice for the monitoring of the degree of contamination of the environment;

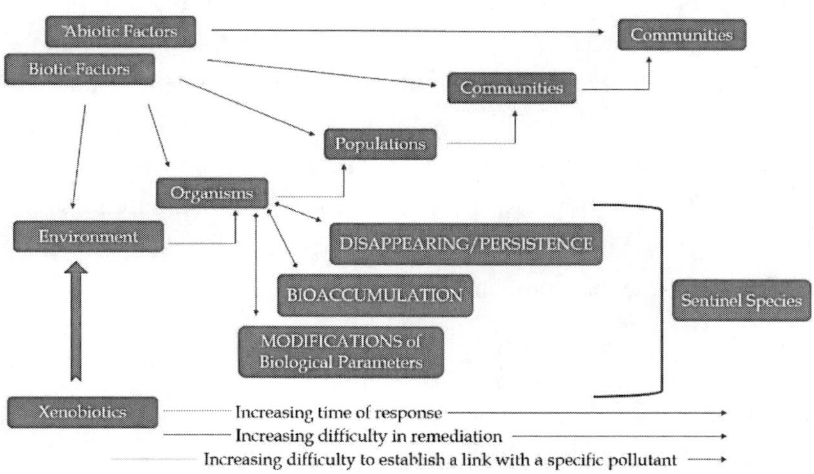

Figure 18. Positions of sentinel species in the various levels of biological organization.

- A species for which modifications of biological parameters at one or various levels of organization (molecular, cellular, physiological, organism, or behavioral) are used to estimate the risks associated with the presence of contaminants.

So, the terms of *bioindicators* and *bioaccumulators* are often used as synonyms for sentinel species. However, although the two terms are indeed associated with the same objective of using living organisms to inform us about the quality of the environment and its consequences, there is a conceptual difference between them.

To avoid problems resulting from different understandings of their separate meanings, and to maintain the integrity of their use, Dauvin and colleagues [484] recommended a clear definition of the meaning of the terms used. It is therefore the definition of a sentinel species as a species showing modifications of biological parameters (indeed biomarkers).

The first sentinel species used were chosen on the criteria of convenience of easy collection, availability, ease of breeding in the laboratory, and good genetic stability, but without ecological relevance. An example is the freshwater crustacean *Daphnia pulex*, widely used to test the toxicity of effluents in the 1970s. Chapman and colleagues [485] showed that the tests carried out on the reproduction of this species showed it to be more sensitive to pollutants than other species of daphnids. Nevertheless, *D. pulex* has been replaced little by little by *Daphnia magna* or *Ceriodaphnia dubia*, both of which are easier to breed in the laboratory.

A second criterion often advanced to justify the choice of a species as a sentinel species is its ease of collection in the field and subsequent manipulation in the laboratory.

Mussels (bivalve mollusks of the family *Mytilidae*) are typical examples, with moreover a large power of bioaccumulation of pollutants that have caused them to be widely used in biomonitoring programs of coastal zones such as "Mussel Watch" in the United States and in the Netherlands.

Furthermore, the biology of mussels, as well as their mechanisms of biotransformation, and the mechanisms of toxicity of xenobiotics, are well known, allowing a good understanding of the fluctuations observed in the biomarkers measured in them [486]. However, several authors [487, 488, 489, 490] have expressed regret that the criteria of selection of these species do not integrate enough of an ecological dimension.

More recently, De Lange and colleagues [491] proposed the assessment of ecosystem vulnerability by using biological and ecological information at different hierarchical levels, in order to bring more ecology into environmental risk assessment, indeed into ecotoxicology, a view supported by [492].

In 2001, Chapman [493] showed the utility and relevance of aquatic oligochaetes in ecological risk assessment, but he noted that, because their taxonomy is difficult and uncertain, they have largely been neglected or mistakenly considered as representing only pollution tolerant taxa. Nevertheless, more recently, Tixier and colleagues [494] indicated that now their taxonomy and autecology are sufficiently known to allow identification to species level and that, when considered at the species level, the oligochaete community represents a highly diverse group with a wide range of ecological roles and a wide range of sensitivity to pollution. Thus, they proposed an approach of sediment quality assessment based on oligochaete community analysis to evaluate the ecological status of storm water ponds.

Furthermore, as pointed out by Lam and Gray [495], it is not uncommon that in the laboratory, the choice of the species to be tested falls on a particular genotype of the species, whereas in the natural environment populations are made up from a mixture of genotypes. These same authors also expressed doubts on the choice of species that are easy to breed and manipulate in the laboratory, considering that such species are the very species that are often intrinsically strongly tolerant to any type of stress, natural or resulting from contamination.

Athrey and colleagues [35] tested the effect of such reduced genetic variation in the least killifish (*Heterandria Formosa*). They used laboratory populations from an eight-generation experiment designed to study selection for resistance to Cd, compared to a control laboratory population and to fish from the original source population. The authors drew two conclusions from this study. The first one is that the loss of genetic variation in genes not directly under selection means that populations that have managed to adapt genetically to an environmental pollutant might be less able to adapt to other environmental stressors. Athrey and colleagues [35] underlined that this could be exacerbated by negative effects on fitness-associated characteristics that are sometimes associated with contaminant adaptation. The second conclusion is that the loss of genetic variation in their laboratory lines appeared to be due to effective population sizes being much lower than actual population sizes, pointing out the necessity of maintaining large laboratory population sizes of organisms used in environmental toxicology studies.

4.1. What Are the Criteria for Selection of Sentinel Species?

In order to reflect the intensity of contaminant exposures, to integrate the effects of the various parameters acting in the environment, and to be able to determine the effects of exposures that may be of weak intensity but endure over a long period, specimens of sentinel species collected in the field must be representative of the local environment and meet several conditions. However, in certain cases, specimens may also be transplanted into the habitat of interest, and maintained at the site in enclosures, active biomonitoring (ABM), in order to follow any change (increase or decrease) in contaminant exposure, and the associated impacts on the individuals, limiting any effects of interspecific biotic factors that might affect interpretation.

Sentinel species have been sought in all environments. Numerous authors [496, 497, 498] have drawn up lists of the characteristics considered to be of greatest importance in these species, but there have been different interpretations, in particular, according to the definition used for a sentinel species and the kind of exposure (in laboratory or in field).

The most usually quoted desirable characteristics are those discussed below:

- Their sedentarity in the site studied, so that the condition of the individuals can be directly correlated to the level of pollution at that site. Animals moving over long distances or visiting the studied sites only occasionally cannot specifically reflect the level of contamination at the site of capture. In this respect, sessile or weakly mobile animals are thus preferred.
- Their ease of identification, capture, or collection all year round, requiring a good knowledge of the ecology of the species and its adaptations to the spatiotemporal variations of the environmental factors in estuarine and coastal zones (temperature, salinity) or the nature of any reproductive migrations.
- A sufficient population size so that any impact of regular collections on population structure and density is negligible. Furthermore, the size of the population available for collection conditions the speed of collecting, thereby affecting the feasibility and the cost of the monitoring program.
- A wide and known distribution range that not only allows comparisons between various sites, but also increases the probability of finding sites that are as weakly impacted as possible by the pollution. However, since it is nearly impossible to find totally cosmopolitan species, closely related species, whether of local or other origin, can be used. Various species of mussels (genera *Mytilus* and *Perna*) or oysters (genera *Crassostrea* and *Saccostrea*) have been preferred in monitoring programs in USA and Europe. Other closely related taxa have also been used to supplement mussels and oysters, such as various species of zebra mussels (genus *Dreissena*) in the region of the Great Lakes, or another bivalve, *Chama sinuata*, in the Florida Keys [499]. Similarly, oysters (*Crassostrea gigas*) and mussels (*Mytilus* spp.) have been used since 1979 for the monitoring of French coasts, whereas similar programs have been set up in numerous other coastal areas.
- A longevity of several years and a resistance to pollutants in the environment, allowing individual specimens to be in contact with

contaminants over long enough periods for any changes in biomarkers to be measurable. This persistence of individuals in the environment also allows changes in parameters to be studied over an extended period, and possibly to reveal any reversibility of ecotoxicological damage if measures have been taken to reduce the level of contaminant exposure. However, the majority of the relevant species have a much lower longevity than humans, and the disorders caused by pollutants appear more quickly, without the latency periods observed in longer-lived animals.

- Dose–effect and cause–effect relationships must be consistently established. Primarily, the species has to present a clear response (effect) to a measurable increase in the level of contaminant (dose). The sentinel species, although resistant to contaminants, does nevertheless have to show sufficient sensitivity to these substances that dysfunctions are apparent, and that these are measurable by techniques in current use. This relationship is often difficult to establish, because of distance to the contaminant source, and of interactions with other stress factors. Responses of the organisms to the pollutants must be early enough to reveal any potential degradation as soon as it occurs, so that substantial ecological damage can be avoided and before any remediation becomes extremely problematic.
- The biology of the species has to be sufficiently well known to differentiate the signal from the background noise. Indeed, intraspecific variability of the biological responses used as biomarkers has been often observed, making the interpretation of data difficult. Furthermore, it is important to be able to make use of control organisms held in parallel, in order to differentiate the changes caused by the contaminants from the natural fluctuations (seasonal, interannual) of the parameter studied. Finally, the characterization of the individuals must be perfectly mastered if the biomarkers to be studied show variation according to sex, age, etc.

The interest of the general public in particular species is also a criterion sometimes necessarily taken into account for the selection of sentinel species. This interest raises new criteria such as recreational value (fishing of salmonids, sports fishing of swordfish), economic value (commercial species), protection status of the species, and any role as flagship species as in the cases of charismatic megafauna such as dugongs, tortoises, seals and sea lions, whales and dolphins, or polar bears.

4.2. Active Biomonitoring

Numerous biotic factors (age, size and weight, sex, stage of sexual maturity) are at the origin of any increase in the intraspecific variability of the responses of sentinel species to contaminant exposure. In order to work on statistically homogeneous populations and to try to be free from this intraspecific variability encountered with individuals taken in their natural environment, ABM by transplantation can be used [500, 501, 502, 503, 504].

ABM uses a set of organisms of known age, and thus at the same stage of sexual maturity, homogeneous in size and weight. These organisms maintained in enclosures are then placed in sites of particular interest. Besides promoting a reduced variability in the responses of the organisms, this methodology also allows control of the period of exposure to contaminants and, at least in theory, to have individuals available for easy sampling at designated times.

ABM operations particularly allow the comparison of various sites, and the distribution of a single species in contaminated sites and/or in control ones. ABM has been extensively used in environmental toxicology, and has been shown to give clearer results than passive biomonitoring since organisms already present *in situ* may have adapted to the local pollutants.

In aquatic ecosystems, transplanting mollusks and fish from a reference site to a polluted area is a feasible strategy for biomonitoring the effects of environmental changes. For example, this technique has been used since 1996 in the Biological Integrators Network on the Mediterranean coast [505].

The use of transplants can only be envisaged if the conditions of the environment allow the species to acclimatize. Conversely, this technique can be also used to try to explain the causes of the absence of a species in an environment, the ecological conditions of which seem nevertheless to be adequate to meet its requirements, or to eliminate some hypotheses on such absences.

Geffard and colleagues [506] showed that the metal contamination of the estuary could not be the cause of this absence, the transplanted mussels developing there as well as oysters, with effective mechanisms of metal detoxification and without deleterious effects on their condition indices.

4.2.1. Risk of Invasive Species

For commercial (furs, food products, etc.) and decorative purposes (stretch of water), in leisure activities (ornamental fish keeping, hunts, fishing) or as subjects of study, man has voluntarily introduced into his environment species that do not usually live there. Few of those succeed in prospering in this new environment, but those that do succeed can be the cause of real ecological disasters. By definition, transplanted sentinel species are able to acclimatize in their new environment, and thus to colonize it. Before deciding to introduce nonnative species in any ABM program, it is thus important to think about the consequences. So, in the example presented above Geffard and colleagues [506], the transplantation of mussels into the Gironde estuary appears to have been without consequences, because the mussels naturally have the opportunity of settling there, but do not so.

4.2.2. Life History

An important factor to be taken into account in the use of sentinel species is the exposure history of the transplanted individuals. In this respect, it is important to pay attention to the history of exposure to pollutants, but also to natural ecological parameters such as bathymetric level on the subtidal zone.

It has been well established for several decades that, within the same species, populations previously exposed to anthropogenic compounds are able to cope with pollutants more efficiently than "naive" individuals.

Takamura and colleagues [507] compared the responses of 118 strains of benthic cyanophyceae and eukaryotic algae, from rivers that are unpolluted or polluted by mining activities (Cu and Zn), on exposure to Cu by measuring their photosynthetic activity. They showed that the exposure concentration of Cu that caused 50% inhibition of photosynthesis (IC_{50}) was positively correlated with the Cu concentrations in the river from which the algae had been isolated.

Berthet and colleagues [508] showed that for identical contaminations with silver, carried out at the same season but with animals of different origins, oysters (*Crassostrea gigas*) of the same size and at the same stage of sexual maturity, could accumulate the metal in quantities varying by one order of magnitude.

Furthermore, bioaccumulation at different levels was associated with different fractionations of silver between the cytosolic and particulate forms of storage in the oysters. These differences were attributed to different genetic characteristics having led to different responses of the organisms in the presence of contaminants.

Furthermore, Wallner-Kersanach and colleagues [509] showed that the rate of elimination of copper in oysters is influenced by their environmental history: the metal is more easily eliminated (30% after 30 days) in animals temporarily (60 days) transplanted into a copper-rich site, than in those which had permanently resided in the same contaminated site (elimination limited to 9% after 30 days). The importance of exposure history was also revealed by Nigro and colleagues [510] and by Regoli and Principato [511] studying, respectively, the differences in the sensitivity of genotoxic and cellular responses, and in the contents of glutathione and of some enzymes involved in the antioxidative defenses, of mussels native to a polluted zone or transplanted from a control zone.

For early and very sensitive biomarkers (e.g., lysosomal membrane integrity and DNA strand breaks), the origin of the mussels had no influence, whereas for biomarkers of chronic contamination (the appearance of micronuclei or lipofuscin accumulation) a higher sensitivity was observed in specimens originating from the contaminated site [510]. Regoli and Principato [511] did not observe differences between mussels subjected to a chronic exposure (natives) and those transplanted (acute exposure) in glutathione levels or in glyoxalase I activity, whereas different activities were observed for catalase and alkaline phosphatase between the two mussel groups.

Bolognesi and colleagues [512] noted that a parallel analysis of caged and wild mussels may improve the characterization of the studied area: wild mussels express higher genetic damage (micronuclei frequency) as a result of cumulative exposure, whereas, in contrast, caged mussels show higher DNA strand breaks as a result of recent exposure.

Chronic exposure to contaminants can lead to the acquisition of a certain tolerance in individuals [6]. Natural selection tends to favor these individuals to the detriment of the ones that are more sensitive to the pollutants, and leads to an increase of the representation of resistance genes over successive generations. Thus, the environmental history of the parents must also be taken into account, and not only that of the animals intended for transplantation. This aspect can be illustrated by the work of Damiens and colleagues [513], who studied the effects of copper contamination on some biomarkers, concentrations of metallothioneins (MT) and thiobarbituric acid reactive substances, catalase (CAT) activity, in oyster larvae (*Crassostrea gigas*) obtained from parents that lived in areas differentially contaminated by this metal. They concluded that larvae of oysters from a strongly copper-contaminated estuary (Bidassoa) are better protected against the toxicity of the metal than the larvae of oysters living in an environment with low copper exposure (Bassin d'Arcachon).

Collin and colleagues [514] exposed to pesticides (lindane, metolachlor, carbofuran), at environmentally relevant concentrations, two populations of oysters from sites differing in pollution levels, particularly pesticide exposure. They observed significant differential gene and protein expression between the two populations, and concluded that—as gene expression is often heritable within populations and supposing that the different sites oyster populations have different genetic backgrounds—these differences could be explained by a combination of environmental and genetic components.

Consequently, whenever it is possible, it is recommended, for bivalves in particular, to use farmed (mariculture) animals for transplantation. For species for which such breeding sources are unavailable, particular attention must be paid to the absence of pollution in the site of origin of the transplants.

If exposure history in terms of exposure to pollutants influences future responses of the organisms, it is also important to pay attention to history of exposure to other ecological parameters that may act as confounding factors. Camus and colleagues [515] showed, for example, that according to their bathymetric position intertidally or subtidally, mussels are more or less accustomed to periods of emersion and immersion, and then react differently to the oxidizing stress generated by the metabolization of pollutants, or simply resulting from collection and transport during the transplantation process.

More generally, Honkoop and colleagues [516] have tested the impact on their physiology (scope for growth) of the stress of handling animals (i.e., mussels), moving them out of their habitat to a new habitat, and caging them. They have used combinations of treatment of mussels including reciprocal transplants, but also false ones consisting of moving mussels but putting them back into their place of origin or in zones of the same contaminant exposure. They concluded that to allow for the confounding effects of moving and disturbance on the physiology of the animals, correct control treatments should be included in the study.

4.2.3. Period of Latency

The use of transplanted sentinel organisms involves a period of latency between the arrival of the individuals in the environment and the beginning of sampling. According to Couillard and colleagues [517] or Geffard and colleagues [518], several months are necessary for the organisms (bivalves in both cases) to reach a state of equilibrium with their environment in terms of their total accumulated metal concentration, and changes over time are also observed in the physicochemical forms of storage of the accumulated metals [519, 520, 521].

Using double transplantations of mussels (*Mytilus galloprovincialis*) between a polluted harbor area and in a weakly impacted site, Regoli and colleagues [522] were able to associate several biomarkers of oxidative stress [antioxidants, total oxyradical scavenging capacity (TOSC)] and changes in cellular structures (integrity of the DNA, lysosomal stability), and observed that not all have the same periods of change, and that the transplantation itself causes an antioxidant response in the initial phase of the experiment.

Camus and colleagues [523], carrying out the same type of experiment to estimate the impacts of the anthropological activities in the Venice urban zone and in various points of Venice lagoon, also demonstrated the influence of transplantation on biomarkers of oxidative stress. Camus and colleagues [523] indicated that 2 weeks are necessary to observe significant differences in TOSC between mussels transplanted into the urban zone and those at the control site, whereas the return to the initial values in the control group is realized only after 4 weeks.

4.3. Concept of Ecosystem Approach

The selection of a sentinel species has to be justified by a recognized link to the structure or the functioning of the ecosystem under surveillance, so that its use brings "a part of the complete puzzle," integrating the interrelations between the various constituents of the studied system. In recent years, several authors [504, 505, 506, 507, 524] have expressed regret that the selection of sentinel species has been based more on anthropomorphic criteria than ecological ones.

For Chapman [504], the species used have too often been selected for their ease of collection and manipulation or breeding in laboratory, whereas their roles in the ecosystem (place in the food chain, the structure of the community, and their functions in the ecosystem) have been neglected. In order to introduce more ECO into ECOtoxicological programs, in natural environments as in the laboratory, it is advisable to study biomarkers in sentinel species that are more representative of the ecosystem. However, if this ecological approach is indispensable, it is not sufficient as pointed out by Chapman [504], who recommends focus on keystone species.

Species interact with numerous constituents of the ecosystem. Where there are no ecological equivalents present, the disturbance of a species, such as by contaminant exposure, can cause deleterious modifications of the quantitative structure of the community, perturb the spatial organization of the community or temporal successions therein. Such species are therefore keystone species. The examples that follow illustrate some levels of actions of these keystone species.

Some species have a particular importance in an ecosystem through their activities of predation or by the biomass that they themselves offer as prey. Among the approximately 3000 species of fishes present in coral reefs, which are among the most biologically productive and diverse ecosystems in the world and among the world's most fragile and endangered ecosystems, the humphead parrotfish (*Bolbometopon muricatum*) is the only predator of the coral whose action can be considered significant. By its bioerosion, it intervenes in the growth of reefs and their spatial distribution, on the shape of coral colonies and their fitness, and on their stability in case of storms or of cyclones [525]. The humphead parrotfish is also a major factor in the redistribution of carbonates in the ecosystem. According to these authors, its depletion, because of pollution and overexploitation is one of the reasons that can explain the regression of coral reefs and the observed changes in the structure of coral communities. These same authors are also concerned, in the longer term, about consequences, which could be dramatic, of the accumulation of carbonates in the reef system.

4.4. Conservation of Biodiversity

In this respect, two questions in particular arise: the protection of the most sensitive species, and the extrapolation of the data acquired on keystone species to all the species constituting the ecosystem. A commonly held proposition is that the protection of the most sensitive species, or of their most sensitive biological stages, in an ecosystem ensures the protection of the whole community.

The validation of a sentinel species is based on the possibility of extrapolating the results obtained on such a species to all the species of the ecosystem. However, even between very closely related taxa, sensitivity to xenobiotics and other contaminants can vary by several orders of magnitude, depending on numerous factors, such as the taxonomic group or, within the same group, a different natural history.

Species that are relatively sensitive or, contrastingly, insensitive to a toxic contaminant are recognized in numerous taxonomic groups, from microorganisms to invertebrates and vertebrates, from microalgae to higher plants, as indeed are contaminant-tolerant populations within the same species.

Bielmyer and colleagues [526] exposed three species of coral to dissolved copper (ranging from 2 to 20 $\mu g\ L^{-1}$) for 5 weeks: the western Atlantic species *Acropora cervicornis* (a threatened species of staghorn coral), *Montastraea faveolata* (star coral), and the Indo-Pacific species *Pocillopora damicornis* (cauliflower coral), each containing different algal symbionts. The three coral species exhibited significantly different sensitivities to copper, with toxic effects occurring in *A. cervicornis* and *P. damicornis* at copper concentrations as low as 4 $\mu g\ L^{-1}$. Copper exposure affected zooxanthellar photosynthesis in *A. cervicornis* and *P. damicornis*, and carbonic anhydrase was significantly decreased in *A. cervicornis* and *M. faveolata*. Likewise, significant decreases in skeletal growth were observed in *A. cervicornis* and *P. damicornis* after copper exposure.

De Pirro and Marshall [527] compared the physiological and mortality responses to copper (Cu^{2+}) of limpets from two different gastropod mollusk lineages: the Prosobranchia (Patellogastropoda: *Helcion concolor* and

Cellana capensis) and the Pulmonata (*Siphonaria serrata* and *Siphonaria capensis*). The *Siphonaria limpets* were nearly one order of magnitude more insensitive to copper than the patellogastropod limpets. Moreover, copper exposure (0.25 ppm Cu^{2+} for 2 h) induced heart rate depression in Siphonaria spp. (to around 50% of the baseline rate), whereas their tissue copper concentrations remained at relatively high control levels. Copper exposure (0.25 ppm Cu^{2+} for 2 h) had no effect on heart rate of the patellogastropod limpets, but led to a significant increase in tissue copper. These authors suggest that the better regulation of internal copper levels by Siphonaria may be related to their exclusive possession of the Cu-requiring respiratory protein hemocyanin in the blood, and that a dependence on relatively high aerobic metabolism by the patellogastropod limpets would limit their capacity for isolation and pollutant avoidance.

Responses to exposure to dissolved silver were compared between four filterfeeding bivalve mollusks, closely related phylogenetically and sharing a common mode of feeding. Sensitive species (a scallop *Chlamys varia* and a mussel *Mytilus galloprovincialis*) and more insensitive ones (an oyster *Crassostrea gigas* and a zebra mussel *Dreissena polymorpha*) were discerned.

Toxic impairments were very variable according to the species: respiration in the mussel, byssus secretion in the scallop, transitory energy impairment in the oyster. Relative sensitivity to silver exposure was not based on an interspecific difference in physiological patterns of silver bioaccumulation, because the (sensitive) scallop and the (insensitive) oyster are strong bioaccumulators, whereas both mussels (one is sensitive, the other one tolerant) are poor bioaccumulators of silver [508].

It is often considered that interspecific variability of sensitivity to contaminants is connected to the very high variability of the physiological adaptive strategies of invertebrates. In the vertebrates with more developed homeostatic processes, a higher homogeneity of responses might therefore be expected. In fact, for four species of marine teleost fish, interspecific differences in behavior were shown in response to the same sublethal dose of copper.

These differences were attributed to a differential sensitivity of the neurophysiological systems between the species, or to differences in the rate of uptake of the pollutant in connection with the relative surface of the body and the gills, knowing that copper and other metals can affect the coat of mucus present on the skin and gills of fish [528]. Kalman and colleagues [529] compared the toxicity of Cd (intraperitoneal injection of 2.5 mg $CdCl_2$ kg^{-1} body weight) in two commercial fish species, *Sparus aurata* and *Solea senegalensis*. These authors showed that metallothionein-like protein (MTLP) synthesis in *S. aurata* probably led to an enhanced ability to cope with the toxicant, whereas MTLP induction in *S. senegalensis* was inhibited by the accumulated cadmium, as reflected by the mortality rate of this species; the authors suggest that MTLPs did not play a main role in metal detoxification by metal sequestration in this latter species.

Eyckmans and colleagues [530] have investigated differences in a Cu-induced oxidative stress response between three freshwater fish species [rainbow trout (*Oncorhynchus mykiss*), common carp (*Cyprinus carpio*), and gibel carp (*Carassius auratus gibelio*)]. They exposed fishes to sublethal Cu concentrations and determined antioxidative enzyme (superoxide dismutase, glutathione reductase and CAT) activities and antioxidant (reduced glutathione and reduced ascorbate) concentrations in gill samples collected from 1 h to 1 month of Cu exposure. They concluded that rainbow trout might use their high reduced GSH concentrations as a first defense against metal exposure, whereas gibel carp and common carp rely more on MT in combination with antioxidant enzymes.

In fishes, interspecific differences of toxicity of organophosphate pesticides can exceed one to two orders of magnitude. Differences of toxicity of diazinon, for example, may be explained by differences in the features of the target enzyme [acetylcholinesterase (AChE)] and in the metabolization of the compound in the liver [531].

Studying the response of brain AChE to methyl paraoxon in more than 20 species of marine fishes, Oliveira and colleagues [532] suggest a possible linkage between brain AChE activity levels and the phylogenetic origin of the fish. In order to use AChE as biomarker for pesticides in the marine environment, these authors recommend as choices as sentinel species those fish that are among the most pesticide-sensitive, which have the highest levels of brain AChE.

As shown by the previous various examples, biological communities are composed of different species, and the inherent sensitivity of each is highly variable. In a contaminated environment, the most sensitive organisms are lost, whereas relatively insensitive or tolerant ones survive and are able to reproduce, leading to a new community that is—as a whole—more tolerant to the contaminant concerned.

In situ, numerous studies carried out on oil spills have shown that crustaceans, in particular amphipods, are very sensitive to hydrocarbons, whereas their impact on polychaetes is much more limited. In a similar manner, copepods seem to be more sensitive to hydrocarbons (and also to anoxia) than nematodes [533]. However, within these groups, an increase in the abundance of opportunist organisms (opportunist polychaetes, oligochaetes, and nematodes) has been observed, in addition to a diminution and a low abundance of the species considered sensitive to the pollution [533].

Similar observations have been reported by Wentsel and colleagues [534], who showed that the populations of oligochaetes increase in lacustrine sediments of contaminated areas (Cd, Zn, and Cr), whereas on the contrary those of chironomids disappear in these same sites. In contrast, the populations of oligochaetes are minimal in sediments in this same lake that are slightly or not contaminated, whereas chironomid populations here are at their greatest.

In response to the European Water Framework Directive frame (EC 2000) on the use of indices and benthic bioindicators to estimate the quality of coastal and estuarine waters, various indices based on differences in sensitivity between species and species succession along a gradient of organic pollution have been developed. Dauvin and Ruellet [535] proposed an index based on the ratio of the opportunist polychaetes to amphipods (benthic opportunistic polychaetes/amphipods). This index can also be used in the case of pollution by hydrocarbons. Recently, this ratio has been tested successfully by Andrade and Renaud [536] along the Norwegian continental shelf for offshore oil and gas impacts.

The reproductive and young growth phases of organisms attract the attention of ecotoxicologists, who consider that they are the most sensitive stages, and therefore key to the preservation of the species in the ecosystem.

Pessoa and colleagues [537] exposed early life stages of tilapia (*Oreochromis niloticus*) at 9 days post-hatch in semistatic experiments to sublethal concentrations of the carbamate insecticide carbofuran (8.3–397 $\mu g\ L^{-1}$). Their results showed effects on cholinesterase activity and a suite of behavioral endpoints (vision, growth and predator avoidance skills at environmentally significant concentrations, and effects on swimming and prey capture only at relatively high concentrations) that are directly related to important ecological mechanisms relevant for their survival, growth, and recruitment to the adult population.

In order to test the total toxicity of dredging materials, Alzieu and colleagues [538] considered bioassays of embryotoxicity in the oyster (*Crassostrea gigas*) or tests on copepodites (juvenile stages) of the copepod crustacean (*Tigriopus brevicornis*) to be sensitive and discriminating, but reported young sea bass or the nauplius larvae of brine shrimp (*Artemia*) to show low sensitivity.

Other authors have also demonstrated that larvae or young do not systematically always show the greatest sensitivities to pollutants. For example, Lotufo and Fleeger [539] exposed two species of meiobenthic copepods (*Schizopera knabeni* and *Nitocra lacustris*) to phenanthrene, and observed that, whereas for the first species the nauplius stage is effectively the most sensitive, for the second, it is the mature females which present a higher sensitivity than all other stages.

Stubblefield and colleagues [540] showed that for rainbow trout (*Oncorhynchus mykiss*) the young were approximately three times less sensitive to the toxic effects of zinc and cadmium than the adults. Furthermore, these authors indicated that the acclimatization of young rainbow trout to sublethal concentrations of cadmium increases their tolerance by a factor of 25, whereas this tolerance is increased only by 4.8 for the adults.

Moreover, as demonstrated by Kammenga and Riksen [541], the intensity of the response of a population to xenobiotics is not always determined by the most sensitive biological stage. Indeed, exposing two species of nematodes to copper, they showed that the population growth rate of the two species is reduced in a similar manner, whereas one of these species shows a higher sensitivity for other life history traits (juvenile survival, duration of the juvenile period and of the period of reproduction, and the daily rate of reproduction).

4.5. Origins of Sentinel Species

Severla authors [6] have reviewed the sources of intraspecific variability of tolerance in sentinel species exposed to contaminants, which may lead to differences in survival of organisms. They discussed the use of wild populations versus "laboratory rats," of populations taken in stable environments *versus* fluctuating ones and of naive populations *versus* exposed ones. Concerning the first point, it is often considered that the genetic impoverishment of laboratory populations is an important *bias* in ecotoxicological tests. According to the origin of the strain (a relatively clean or polluted site), the clones used in ecotoxicological bioassays may vary in their tolerance to xenobiotics and other contaminants, and the results obtained may thus lead to under- or overestimation of the risk for a natural population. This can be taken into account by the application of safety factors. On the second point, in laboratory studies, test organisms are almost always kept in optimal conditions of temperature, oxygen, salinity, food, etc., in order to optimize performance in the control, and to isolate the effects of the tested chemical. On the contrary, in their natural environment, organisms are rarely in optimal conditions, and have to cope with environmental stresses.

Thus, for a long time, euryoecious species, such as estuarine organisms, have been considered to be more tolerant to xenobiotics than stenoecious ones, but this assertion has still not been verified [6], and that a substantial part of uncertainty in toxicity tests could be explained by natural factors.

The mechanisms that allow organisms to cope with the presence of various contaminants in their environment have been recently reviewed by

Johnston and Rainbow and Luoma for metal tolerance, for antioxidant defenses, for biotransformation of organic contaminants, and for stress proteins [6]. They may be individual physiological acclimations after exposure to a contaminant (mechanisms of defense); in this case, the tolerance may be lost if exposures are not maintained, or genetic adaptations (heritable) when a selective force applied by a contaminant is maintained across multiple generations.

In all cases, both inter- and intraspecific variabilities of sensitivity of organisms to pollutants have to be taken into account in order to correctly interpret the results of toxicity tests, and to avoid, as much as possible, any under- or overestimation of risk.

The ecotoxicological endpoint measured, and the type of test used, several species and endpoints have to be investigated in risk assessment, in order to reach a sound extrapolation at the level of the whole ecosystem, because the sensitivity of a given species depends on the involved pollutant.

4.6. Choice of Physical Compartment

In the aquatic environments, ecotoxicological research has, over a long period, mainly concerned species that are representative of the water column. On the other hand, although sediment quantitatively represents the most contaminated environment, both by organic and inorganic pollutants [542, 543], the endofauna and endoflora were relatively little taken into account during this period.

Moreover, Chapman and Wang [544] underline that most methods of evaluating the toxicities of sediment are especially applicable to freshwater or marine sediments, but few take into account the specific characteristics of estuaries. These authors attempted to produce an inventory of the species used in tests of the overall toxicity of estuarine sediments. They noted a relative phylogenetic poverty, with a wide dominance of crustaceans, in particular amphipods, whereas they wondered at the absence of some taxa such as the aquatic oligochaetes or the chironomids (midge larvae), and at the low representation of fish. In the same work, they regret that too many tested species are more marine than estuarine, and that allowances for salinity levels are insufficient. Authors too often try to fit the salinity of a sediment to the available model species, thereby modifying the physicochemical forms of pollutants and therefore their bioavailability and potential toxicity, rather than testing the relevant sediment while maintaining its initial physicochemical characteristics, and selecting the species to be tested on these criteria.

Chapman and Wang [544] presented the necessity to develop studies of this physical compartment in estuaries, regions that are particularly subjected to anthropogenic contributions, often in complex mixtures, and to pay more attention to the physicochemical conditions of the laboratory tests in order to be able to compare the laboratory results to those in situ with more justification.

In this perspective, few endobenthic estuarine species have been selected as biological models in ecotoxicological programs, typically the ragworm (*Nereis diversicolor*) and the bivalve (*Scrobicularia plana*), which are indeed keystone species of the estuarine ecosystem.

With the aim of a general appreciation of the quality of the estuarine environments, these two deposit feeding estuarine invertebrates appear as an interesting complement to filter feeders representative of the quality of the water column, allowing a wider appreciation of the contamination of sediments *in situ*, and over relatively long periods [545, 546, 547, 548, 549, 550].

Among fishes, the European eel, *Anguilla anguilla*, a bottom-dwelling species has been chosen by Piva and colleagues [551] and Benedetti and colleagues [552] as a sentinel species for assessing sediment quality because of its frequency in European estuaries, brackish and freshwater basins, and its tolerance to a wide range of biotic and abiotic factors. Moreover, ecotoxicological studies have shown their capacity to accumulate elevated concentrations of chemicals due to their high lipid content [553, 554], and several biomarkers have been characterized that respond in this fish to the main classes of environmental pollutants [555, 556].

4.7. USED SPECIES AS SENTINELS

Many species, animal or plant, can be used to provide indications on the quality of the environment, but their performance and hence suitability in this role may well turn out to differ according to the biomarker chosen. This factor, added to the list of criteria already evoked in the choice of a sentinel species, may further quickly limit the choice for a particular objective [557].

In a review on the use of MT as a biomarker, Amiard and colleagues [91] show that for this biomarker, three taxa of invertebrates are mainly used in coastal and estuarine regions, where they play key roles: mollusks, crustaceans, and annelid worms. Among the mollusks, the bivalves resulted to be mostly used in this role (65%), whereas decapods dominate the crustaceans (70%), and polychaetes are the annelids most often quoted (81%).

Although mollusks are often recommended for use to estimate lysosomal membrane integrity, DNA strand breaks, or MT concentrations, they are, on the other hand, of less interest in investigations, for example, of the catalytic activity of P450. Other taxa are then recommended such as particular teleost fishes: red mullet (*Mullus barbatus*) or bass (*Serranus* spp.) [557]. In a recent review, Amiard-Triquet and colleagues [6] have highlighted the importance of polychaetes in ecotoxicological studies, because of the robustness of many polychaete worms, and because many biomarkers measured in these animals have proven to be generally more sensitive to pollution factors than to natural factors (e.g., salinity of the studied area, size of the worms), thereby avoiding many of the problems caused by so-called "confounding factors."

Amiard-Triquet and colleagues [6] conclude that ragworms of the family *Nereidae* appear to represent the most commonly used taxon of polychaetes for the determination of biomarkers. On the other hand, mixed populations may supply different data to those obtained on isolated organisms. Chapman and colleagues [563], using mixed populations of freshwater oligochaetes (*Limnodrilus hoffmeisteri* and *Tubifex tubifex*), demonstrated that mixed populations of these two species showed higher tolerance to cadmium, mercury, and sodium pentachlorophenol, but lower tolerance to anoxia, than isolated populations. Similar conclusions were reached by Keilty and colleagues [564], who showed that, during contamination of sediment by endrin, *L. hoffmeisteri* benefits from the presence of the oligochaete *Stylodrilus heringianus*, with a decrease of its mortality, an increase of its mass, and a stimulation of its bioturbatory activity.

It can be concluded that the selection of a single species as a unique sentinel organism is unrealistic. Thus, the choice has to turn toward a set of species that can represent at best the various functional aspects of an ecosystem. According to the specific objectives under consideration, various approaches (described by Figure 19) are possible:

- In an analysis of the environmental risks associated with the discharge of a xenobiotic contaminant, bioassays can be carried out on various species, according to a methodology approved by national and international authorities such as the European Commission (2000, 2003). After the application of a safety factor, guideline values can be established, fixing quantitative limits not to be exceeded in order to protect the quality of the environment. Nevertheless, the evaluation of risk could benefit from the use of biomarkers in controlled exposures.
- When the objective is the protection of the biodiversity in an area of particular interest, the use of groups of bioindicator species allows a long-term evaluation of contaminant effects on the resident populations and communities. The criteria of absence or presence of sensitive species can be investigated and core biomarkers of stress can be measured in these organisms, but it will remain difficult to establish relationships between cause and effect.
- The objectives of the present work are to develop a biomarker tools using sentinel species recognized as successful in the manifestation of biomarkers revealing contamination of the environment, and the resulting damage to the organisms, that will reverberate at the level of the population, the community, and ultimately, the ecosystem.

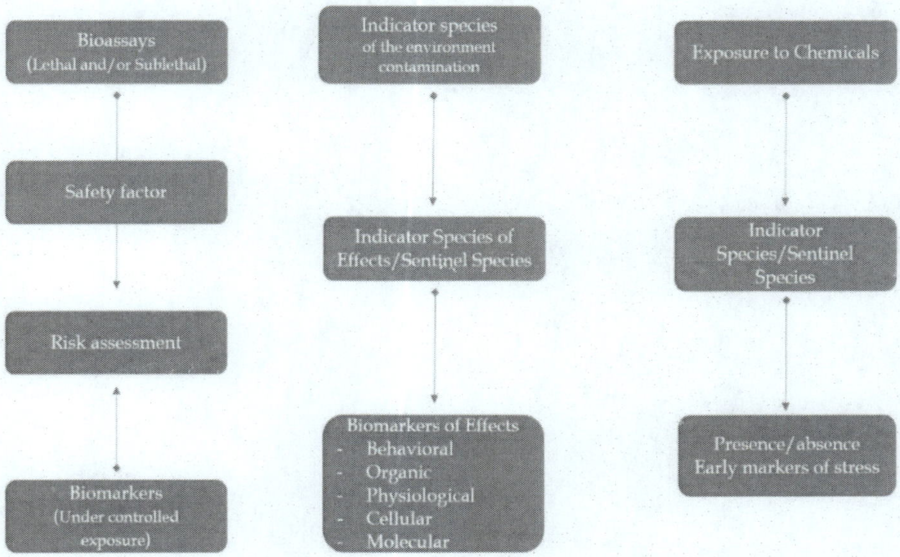

Figure 19. Different uses of sentinel species according to the chosen objectives. Ixtaina, Y., Capitani, M. I., Nolasco, S. M. and Tomás, M. C. (2013). Chia *(Salvia hispanica L.)* as a source of bioactive compounds for food processing. *Oils and Fats*, 92 (III): 410-427. In reverse again, although only few data for glutathione-S-transferase (GST) are supplied by fish [558, 559], GST is successfully used as biomarker of exposure in mollusks [560, 561] and annelids [562, 561].

References

[1] Letcher, R. J., Bustnes, J. O., Dietz R., Jenssen, B. M., Jørgensen, E. H., Sonne, C., Verreault, J., Vijayan, M. M. and Gabrielsen, G. W. (2010). Exposure and effects assessment of persistent organohalogen contaminants in arctic wildlife and fish. *Science of the Total Environment*, 408: 2995–3043.

[2] Mason, A. Z. and Jenkins, J. D. (1995). Metal detoxification in aquatic organisms. In *Metal Speciation and Bioavailability in Aquatic Systems*, ed. A. Tessier and D. R. Turner, 479–608. Chichester: Wiley.

[3] Amiard, J. C. Amiard-Triquet, C., Barka, S., Pellerin, J. and Rainbow, P. S. (2006). Metallothioneins in aquatic invertebrates: Their role in metal detoxification and their use as biomarkers. *Aquatic Toxicology*, 76 (II): 160–202.

[4] Serén, N., Glaberman, S., Carretero, M. A. and Chiari, Y. (2014) Molecular evolution and functional divergence of the metallothionein gene family in vertebrates. *Journal of Molecular Evolution*, 78(III-IV): 217-233.

[5] Marigomez, I., Soto, M., Cajaraville, M. P., Angulo, E. and Giamberini, L. (2002). Cellular and subcellular distribution of metals in molluscs. *Microscopy Research and Technology*, 56 (V): 358–392.

[6] Amiard-Triquet, C., Rainbow, P. S. and Roméo, M. (2011). *Tolerance to Environmental Contaminants.* Boca Raton, FL: CRC Press.

[7] Newman, M. C. and Clements, W. H. (2008). *Ecotoxicology. A Comprehensive Treatment.* Boca Raton, FL: CRC Press.

[8] Anderson, J. W. and Lee, R. F. (2006). Use of biomarkers in oil spill risk assessment in the marine environment. *Human Ecology and Risk Assessment*, 12: 1192–1222.

[9] Thain, J. E., Vethaak, A. D. and Hylland, K. (2008). Contaminants in marine ecosystems: Developing an integrated indicator framework using biological-effect techniques. *ICES Journal of Marine Sciences*, 65: 1508–1514.

[10] Manahan, S. E. (2003). *Toxicological Chemistry and Biochemistry*. Boca Raton, FL: Lewis Publishers.

[11] Kroon, F., Streten, C. and Harries, S. (2017) A protocol for identifying suitable biomarkers to assess fish health: A systematic review. *PLoS ONE*, 12(IV): e0174762.

[12] Chapman, P. M. (2002). Integrating toxicology and ecology: Putting the "eco" into ecotoxicology. *Marine Pollution Bulletin*, 44: 7–15.

[13] Kammenga, J. E. and Riksen J. A. G. (1996). Comparing differences in species sensitivity to toxicants: Phenotypic plasticity versus concentration–response relationships. *Environmental Toxicology and Chemistry*, 15: 1649–1653.

[14] Hosokawa, Y. (1993). Remediation work for mercury contaminated bay – experiences of Minamata bay project, Japan. *Water Science and Technology*, 28: 339–348.

[15] Tannenbaum, L. V. (2005). A critical assessment of the ecological risk assessment process: A review of misapplied concepts. *Integrated Environmental Assessment and Management*, 1: 66–72.

[16] Sternberg, R. M., Gooding, M. P., Hotchkiss A. K. and Le Blanc, G. A. (2010). Environmental-endocrine control of reproduction in gastropods: Implications for the mechanisms of tributyl-induced imposex in prosobranchs. *Ecotoxicology*, 19: 4–23.

[17] Bryan, G. W., and Gibbs, P. E. (1991). Impact of low concentrations of tributyltin (TBT) on marine organisms: A review. In *Metal ecotoxicology, Concepts and Applications*, ed. M. Newman and A. W. McIntosh, 323–361. Chelsea, MI: Lewis Publishers.

[18] Jugan, M. L. Oziol, L., Bimbot, M,, Huteau, V,, Tamisier-Karolak, S., Blondeau, J. P. and Lévi, Y. (2009). In vitro assessment of thyroid and estrogenic endocrine disruptors in wastewater treatment plants, rivers and drinking water supplies in the greater Paris area (France). *Science of the Total Environment*, 407: 3579–3587.

[19] Bertin, A., Inostroza, P. A. and Quinones, R. (2011). Estrogen pollution in a highly productive ecosystem off central-south Chile. *Marine Pollution Bulletin*, 62: 1530–1537.

[20] Gong, J. Ran, Y., Chen, D. Y. and Yang, Y. (2011). Occurrence of endocrine-disrupting chemicals in riverine sediments from the Pearl River Delta, China. *Marine Pollution Bulletin*, 63: 556–563.

[21] Tetreault, G. R. Bennett, C. J., Shires, K., Knight, B., Servos, M. R. and Mc Master M. E. (2011). Intersex and reproductive impairment of wild fish exposed to multiple municipal wastewater discharges. *Aquatic Toxicology*, 104: 278–290.

[22] Baroiller, J. and D'Cotta, H. (2001). Environment and sex determination in farmed fish. *Comparative Biochemistry Physiology*, 130: 399–409.

[23] Langston, W. J., Burt, G. R. and Chesman, B. S. (2007). Feminisation of male clams *Scrobicularia plana* from estuaries in Southwest UK and its induction by endocrine-disrupting chemicals. *Marine Ecology-Progress Series*, 333: 173–184.

[24] Blanton, M. L. and Specker, J. L. (2007). The hypothalamus–pituitary–thyroid (HPT) axis in fish and its role in fish development and reproduction. *Critical Reviews in Toxicology*, 37: 97–115.

[25] Carr, J. A. and Patiño, R. (2011). The hypothalamus–pituitary–thyroid axis in teleosts and amphibians: Endocrine disruption and its consequences to natural populations. *General and Comparative Endocrinology*, 170: 299–312.

[26] Dell'Omo, G. (2002). *Behavioural Ecotoxicology*. Chichester, UK: Wiley.

[27] Amiard-Triquet, C. and Rainbow, P. S. (2009). Conclusions. In *Environmental Assessment of Estuarine Ecosystems. A Case Study*,

ed. C. Amiard-Triquet and P. S. Rainbow, 323–348. Boca Raton, FL: CRC Press.

[28] Hellou, J. (2011). Behavioural ecotoxicology, an "early warning" signal to assess environmental quality. *Environmental Science and Pollution Research*, 18: 1–11.

[29] Maltby, L., Kedwards, T. K. and Forbes, V. E. (2001). Linking individual-level responses and population-level consequences. In *Ecological variability: Separating Natural from Anthropogenic Causes of Ecosystem Impairment*, ed. D. J. Baird and G. A. Burton, 27–82. Pensacola, FL: Society of Environmental Toxicology and Chemistry (SETAC).

[30] Charles, S., Billoir, E., Lopes, C. and Chaumot, A. (2009). Matrix population models as relevant modelling tools in ecotoxicology. In *Ecotoxicology Modelling*, ed. J. Devillers, 261–298. Dordrecht: Springer.

[31] Peterson, C. L. and Côté, J. (2004). Cellular machineries for chromosomal DNA repair. *Genes and Development*, 18: 602–616.

[32] Hemingway, J., Hawkes, N. J., Mc Carroll, L. and Ranson, H. (2004). The molecular basis of insecticide resistance in mosquitoes. *Insect Biochemistry and Molecular Biology*, 34: 653–665.

[33] Nies, D. H. (1999). Microbial heavy-metal resistance. *Applied Microbiology and Biotechnology*, 51: 730–750.

[34] Nevo, E., Ben-Shlomo, R. and Lavie, B. (1984). Mercury selection of allozymes in marine organisms: Predictions and verification in nature. *Proceedings of the National Academy of Sciences of the United States of America National Academy of Sciences*, 81: 1258–1259.

[35] Athrey, N. R. G., Leberg, P. L. and Klerks P. L. (2007). Laboratory culturing and selection for increased resistance to cadmium reduce genetic variation in the least killifish, *Heterandria formosa*. *Environmental Toxicology and Chemistry*, 26: 1916–1921.

[36] Lagadic, L. (1997). *Biomarqueurs en écotoxicologie. Aspects fondamentaux.* Paris: Masson. [*Biomarkers of ecotoxicology, Fundamental aspects*. Paris: Masson.]

[37] Lagadic, L., Caquet, T., Amiard, J. C. and Ramade, F. (1998). *Utilisation de biomarqueurs pour la surveillance de la qualité de l'environnement.* Paris: Lavoisier Tec & Doc. [*Use of biomarkers for monitoring the environment quality.* Paris: Lavoisier Tec & Doc.]

[38] Garrigues, P., Barth, H., Walker, C. H. and Narbonne, J-F. (2001) *Biomarkers in Marine Organisms: A Practical Approach.* Amsterdam: Elsevier Science.

[39] Allan, I. J., Vrana, B., Greenwood, R., Mills, G. A., Roig, B. and Gonzalez, C. (2006). A "toolbox" for biological and chemical monitoring requirements for the European Union's Water Framework Directive. *Talanta*, 69: 302–322.

[40] Hagger, J. A., Jones, M. B., Lowe, D., Leonard, D. R., Owen, R. and Galloway, T. S. (2008). Application of biomarkers for improving risk assessments of chemicals under the Water Framework Directive: A case study. *Marine Pollution Bulletin*, 56: 1111–1118.

[41] Sanchez, W. and Porcher J. M. (2009). Fish biomarkers for environmental monitoring within the Water Framework Directive of the European Union. *TRAC-Trends and Annals in Chemistry*, 28: 150–158.

[42] He, X., Nie, X., Wang, Z., Cheng, Z., Li, K., Li, G., Hung Wong, M., Liang, X., Tsui, M. T. (2011). Assessment of typical pollutants in waterborne by combining active biomonitoring and integrated biomarkers response. *Chemosphere,* 84: 1422–1431.

[43] Ramos-Gómez, J., Coz, A., Viguri, J. R., Luque, A., Martín-Díaz, M. L. and Del Valls, T. Á. (2011). Biomarker responsiveness in different tissues of caged *Ruditapes philippinarum* and its use within an integrated sediment quality assessment. *Environmental Pollution*, 159: 1914–1922.

[44] Trujillo-Jiménez, P., Elías Sedeño-Díaz, J., Camargo, J. A. and López-López, E. (2011). Assessing environmental conditions of the Río Champotón (México) using diverse indices and biomarkers in the fish *Astyanax aeneus* (Günther, 1860). *Ecological Indicators*, 11: 1636–1646.

[45] Martínez-Gómez, C., Vethaak, A. D., Hylland, K., Burgeot, T., Köhler, A., Lyons, B. P., Thain, J., Gubbins M. J. and Davies I. M. (2010). A guide to toxicity assessment and monitoring effects at lower levels of biological organization following marine oil spills in European waters. *ICES Journal of Marine Sciences*, 67: 1105–1118.

[46] Forbes, V. E., Plamqvist, A. and Bach, L. (2006). The use and misuse of biomarkers in ecotoxicology. *Environmental Toxicology and Chemistry*, 25: 272–280.

[47] Cairns, J. Jr. (1992). The threshold problem in ecotoxicology. *Ecotoxicology*, 1: 3–16.

[48] Kalman, J., Buffet, P. E., Amiard, J. C., Denis, F., Mouneyrac, C. and Amiard-Triquet, C. (2010). Assessment of the influence of confounding factors (weight, salinity) on the response of biomarkers in the estuarine polychaete *Nereis diversicolor*. *Biomarkers*, 15: 462–469.

[49] Amiard-Triquet, C. and Rainbow, P. S. (2009). Conclusions. In *Environmental Assessment of Estuarine Ecosystems. A Case Study*, ed. C. Amiard-Triquet and P. S. Rainbow, 323–348. Boca Raton, FL: CRC Press.

[50] Sanchez, W., Piccini, B., Maillot-Maréchal, E. and Porcher, J. M. (2010). Comparison of two reference systems for biomarker data analysis in a freshwater biomonitoring context. *Environment International*, 36: 377–382.

[51] National Research Council (NRC). (1987). Biological markers in environmental health research. *Environmental Health Perspectives*, 74:3–9.

[52] National Research Council (NRC). (1989). *Biological Markers in Reproductive Toxicology*. Washington, DC: National Academic Press.

[53] McCarthy, J. F. and Shugart, L. R. (1990). *Biomarkers of Environmental Contamination.* Boca Raton, FL: Lewis Publishers.

[54] Depledge, M. H. (1994). The rational basis for the use of biomarkers as ecotoxicological tools. In *Nondestructive Biomarkers in Vertebrates*, ed. M. C. Fossi and C. Leonzio, 271–95. Boca Raton, FL: Lewis Publishers.

[55] Depledge, M. H. and Fossi, M. C. (1994). The role of biomarkers in environmental assessment. *Invertebrates. Ecotoxicology*, 3: 161–72.

[56] Peakall, D. B. and Walker, C. H. (1994). The role of biomarkers in environmental assessment. *Vertebrates. Ecotoxicology*, 3: 173–9.

[57] Ernst, W. H. O., and Peterson, P. J. (1994). The role of biomarkers in environmental assessment. *Terrestrial plants. Ecotoxicology,* 3: 180–92.

[58] Lagadic, L., Caquet, T. and Ramade, F. (1994). The role of biomarkers in environmental assessment. *Invertebrate populations and communities. Ecotoxicology*, 3:193–208.

[59] Van Gestel, C. A. M. and Van Brummelen, T. C. (1996). Incorporation of the biomarker concept in ecotoxicology calls for a redefinition of terms. *Ecotoxicology,* 5: 217–25.

[60] Van der Oost, R., Porte-Visa, C. and Van den Brink, N. W. (2005). Biomarkers in environmental assessment. In *Ecotoxicological Testing of Marine and Freshwater Ecosystems: Emerging Techniques, Trends, and Strategies,* ed. P. J. Den Besten and M. Munawar, 87–152. Boca Raton, FL: Taylor & Francis.

[61] De Lafontaine, Y. Gagné, F., Blaise, C., Costan, G., Gagnon, P. and Chan, H. M. (2000). Biomarkers in zebra mussels (*Dreissena polymorpha*) for the assessment and monitoring of water quality of the St Lawrence River (Canada). *Aquatic Toxicology*, 50: 51–71.

[62] Pampanin, D. M., Anderson, O. K. and Viarengo, A. (2006). Background for the BEEP Stavanger workshops: Biological effects on marine organisms in two common, large, laboratory experiments and in a field study. Comparison of the value (sensitivity, specificity, etc.) of core and new biomarkers. *Aquatic Toxicology*, 78(S): 1–4.

[63] van der Oost, R., Beyer, J. and Vermeulen, N. P. E. (2003) Fish bioaccumulation and biomarkers in environmental risk assessment: a review. *Environmental Toxicology and Pharmacology*, 13: 57-149.

[64] Schlenk, D. (1999). Necessity of Defining Biomarkers for Use in Ecological Risk Assessments. *Marine Pollution Bulletin*, 39: 48-53.

[65] Suter, G. W. I. I. (1993). *Ecological Risk Assessment*. Boca Raton, Florida, USA: Lewis Publishers.

[66] Hook, S. E., Gallagher, E. P. and Batley, G. E. (2014). The Role of Biomarkers in the Assessment of Aquatic Ecosystem Health. *Integrated Environmental Assessment and Management*, 10: 327-341.

[67] Mohammed, A., May, T., Echols, K., Walther, M., Manoo, A. and Maraj, D. (2012). Metals in sediments and fish from Sea Lots and Point Lisas Harbors, Trinidad and Tobago. *Marine Pollution Bulletin*, 64: 169-173.

[68] Chen, M. H. and Chen, C. Y. (1999). Bioaccumulation of sediment-bound heavy metals in grey mullet, Liza macrolepis. *Marine Pollution Bulletin*, 39: 239-244.

[69] Qiu, Y-W. (2015). Bioaccumulation of heavy metals both in wild and mariculture food chains in Daya Bay, South China. *Estuarine, Coastal and Shelf Science*. 163: 7-14.

[70] Fernandes, C., Fontainhas-Fernandes, A., Peixoto, F., Salgado, M. A. (2007). Bioaccumulation of heavy metals in *Liza saliens* from the Esmoriz-Paramos coastal lagoon, Portugal. *Ecotoxicology and Environment Safety*, 66: 426-431.

[71] Zhang, L., Shi, Z., Jiang, Z., Zhang, J., Wang, F. and Huang, X. (2015). Distribution and bioaccumulation of heavy metals in marine organisms in east and west Guangdong coastal regions, South China. *Marine Pollution Bulletin*, 101(II): 930-937.

[72] Waltham, N. J., Teasdale, P. R. and Connolly, R. M. (2011). Contaminants in water, sediment and fish biomonitor species from natural and artificial estuarine habitats along the urbanized Gold Coast, Queensland. *Journal of Environment Monitoring*, 13: 3409-3419.

[73] Chen, M. H. (2002). Baseline metal concentrations in sediments and fish, and the determination of bioindicators in the subtropical Chi-ku Lagoon, SW Taiwan. *Marine Pollution Bulletin*, 44: 703-714.

[74] Beg, M. U., Al-Jandal, N., Al-Subiai, S., Karam, Q., Husain, S. and Butt, S. A. (2015). Metallothionein, oxidative stress and trace metals in gills and liver of demersal and pelagic fish species from Kuwaits' marine area. *Marine Pollution Bulletin,* 100: 662-672.

[75] Souza, I. C., Duarte, I. D., Pimentel, N. Q., Rocha, L. D., Morozesk, M., Bonomo, M. M., (2013). Matching metal pollution with bioavailability, bioaccumulation and biomarkers response in fish (Centropomus parallelus) resident in neotropical estuaries. *Environmental Pollution*, 180: 136-144.

[76] Nel, L., Strydom, N. A. and Bouwman, H. (2015). Preliminary assessment of contaminants in the sediment and organisms of the Swartkops Estuary, South Africa. *Marine Pollution Bulletin*, 101: 878-885.

[77] Qiu, Y-W., Lin, D., Liu, J-Q. and Zeng, E. Y. (2011). Bioaccumulation of trace metals in farmed fish from South China and potential risk assessment. *Ecotoxicological Environmental Safety*, 74: 284-293.

[78] Jeffree, R. A., Warnau, M., Teyssie, J-L. and Markich, S. J. (2006). Comparison of the bioaccumulation from seawater and depuration of heavy metals and radionuclides in the spotted dogfish *Scyliorhinus canicula* (Chondrichthys) and the turbot *Psetta maxima* (Actinopterygii: Teleostei). *Science of the Total Environment*, 368: 839-852.

[79] Benhamed, S., Guardiola, F. A., MartõÂnez, S., MartõÂnez-SaÂnchez, M. J., PeÂrez-Sirvent, C. and Mars, M. (2016). Exposure of the gilthead seabream (*Sparus aurata*) to sediments contaminated with heavy metals downregulates the gene expression of stress biomarkers. *Toxicology Reports*, 3: 364-372.

[80] Dutton, J. and Fisher, N. S. (2010). Intraspecific comparisons of metal bioaccumulation in the juvenile Atlantic silverside *Menidia menidia*. *Aquatic Biology*, 10: 211-226.

[81] Barbee, N. C., Ganio, K. and Swearer, S. E. (2014). Integrating multiple bioassays to detect and assess impacts of sublethal exposure to metal mixtures in an estuarine fish. *Aquatic Toxicology*, 152: 244-255.

[82] Edwards, J. W., Edyvane, K. S., Boxall, V. A., Hamann, M. and Soole, K. L. (2001). Metal levels in seston and marine fish flesh near

industrial and metropolitan centres in South Australia. *Marine Pollution Bulletin*, 42: 389-396.

[83] Mzimela, H. M., Wepener, V. and Cyrus, D. P. (2003). Seasonal variation of selected metals in sediments, water and tissues of the groovy mullet, *Liza dumerelii* (Mugilidae) from the Mhlathuze Estuary, South Africa. *Marine Pollution Bulletin*, 46: 659-664.

[84] Vicente-Martorell, J. J., Galindo-Riano, M. D., Garcia-Vargas, M. and Granado-Castro, M. D. (2009). Bioavailability of heavy metals monitoring water, sediments and fish species from a polluted estuary. *Journal of Hazardous Materials*, 162: 823-836.

[85] Aas, E., Baussant, T., Balk, L., Liewnborg, B. and Andersen, O. K. (2000). PAH metabolites in bile, cytochrome p4501A and DNA adducts as environmental risk parameters for chronic oil exposure: a laboratory experiment with Atlantic cod. *Aquatic Toxicology*, 51: 241–258.

[86] Denslow, N. D., Kochera, J., Sepulveda, M. S., Gross, T. and Holm, S. E. (2004). Gene expression fingerprints of largemouth bass (*Micropterus salmoides*) exposed to pulp and paper mill effluents. *Mutation Research,* 552: 19–34.

[87] Folmar, L. C., Denslow, N. D., Rao, V., Chow, M., Crain, D. A., Enblom J., Marcino, J. and Guillette, L. J. (1996). Vitellogenin induction and reduced serum testosterone concentrations in feral male carp (*Cyorinus carpio*) captured near a major metropolitan sewage treatment plant. *Environmental Health Perspectives*, 104: 1096–1101.

[88] Jobling, S., Nolan, M., Tyler, C. R., Brighty, G., Sumpter, J. P. (1998). Widespread sexual disruption in wild fish. *Environmental Science and Technology*, 32: 2498–2506.

[89] Johnson, L. L., Lomax, D. P., Myers, M. S., Olson, O. P., Sol, S. Y., O'Neill S. M., West, J. and Collier, T, K. (2008). Xenoestrogen exposure and effects in English sole (*Parophyrys vetulus*). *Aquatic Toxicology*, 88: 29–38.

[90] Johnson, L. L., Arkossh, M. R., Bravo, C. F., Collier, T. K., Krahn, M. M., Meador, J. P., Myers, M. S., Reichert, W. L. and Stein, J. E. (2008). The effects of polycyclic aromatic hydrocarbons in fish from

Puget Sound, Washington. In: Di Giulio, RT.; Hinton, DE., editors. *The Toxicology of Fishes.* CRC Press; Boca Raton, FL, USA. p. 877-923.

[91] Amiard, J-C., Amiard-Triquet, C., Barka, S., Pellerin, J. and Rainbow, J. S. (2006). Metallothioneins in aquatic invertebrates: their role in metal detoxification and their use as biomarkers. *Aquatic Toxicology*, 76: 160–202.

[92] Hook, S. E., Skillman, A. D., Small, J. A. and Schultz, I. R. (2006) Gene expression patterns in Rainbow Trout, *Onorhynchus mykiss*, exposed to a suite of model toxicants. *Aquatic Toxicology*, 77: 372–385.

[93] Tom, M., Moran, O., Jakubov, E., Cavar, B. and Rinkevic, B. (1998). Molecular characterization of metallothionein-cDNA of Sparus aurata used for detecting heavy metal pollution along the Mediterranean coast of Israel. *Marine Pollution Bulletin*, 36: 131–137.

[94] Chesman, B. S., O'Hara, S., Burt, G. R. and Langston, W. J. (2007). Hepatic metallotheionein and total oxyradical scavenging capacity in Atlantic cod *Gadus morhua* caged in open sea contamination gradients. *Aquatic Toxicology*, 84: 310–320.

[95] Williams, C. R. and Gallagher, E. P. (2013) Effects of cadmium on olfactory mediated behaviors and molecular biomarkers in coho salmon (*Oncorhynchus kisutch*). *Aquatic Toxicology*, 141(C): 295–302.

[96] Klumpp, D. W., Humphrey, C., Huasheng, H. and Tao, F. (2002). Toxic contaminants and their biological effects in coastal waters of Xiamen, China. II. Biomarkers and embryo malformation rates as indicators of pollution stress in fish. *Marine Pollution Bulletin*, 44: 761–769.

[97] Cooper, N. L. and Bidwell, J. R. (2006). Cholinesterase inhibition and impacts on behavior of the Asian clam, *Corbicula fluminea*, after exposure to an organophosphate insecticide. *Aquatic Toxicology*, 76: 258–267.

[98] Laetz, C. A., Baldwin, D. A., Collier, T. K., Herbert, V., Stark, J. D. and Schol, N. L. (2009). The synergistic toxicity of pesticide mixtures:

implications for risk assessment and the conservation of endangered Pacific salmon. *Environmental Health Perspectives*, 117:348–353.

[99] Annabi, A., Kessabi, K., Navarro, A., Said, K., Messaoudi, I. and Pina, B. (2012). Assessment of reproductive stress in natural populations of the fish *Aphanius fasciatus* using quantitative mRNA markers. *Aquatic Biology*, 17: 285-288.

[100] Townsend, D. M. and Tew, K. D. (2003). The role of glutathione-S-transferase in anti-cancer drug resistance. *Oncogene* 22: 7369-7375.

[101] Omar, W. A., Saleh, Y. S. and Marie, M-AS. (2016). The use of biotic and abiotic components of Red Sea coastal areas as indicators of ecosystem health. *Ecotoxicology*, 25: 253-266.

[102] Fasulo, S., Mauceri, A., Maisano, M., Giannetto, A., Parrino, V. and Gennuso F. (2010). Immunohistochemical and molecular biomarkers in *Coris julis* exposed to environmental contaminants. *Ecotoxicology and Environmental Safety*, 73: 873-882.

[103] De Domenico, E., Mauceri, A., Giordano, D., Maisano, M., Giannetto, A. and Parrino, V. (2013). Biological responses of juvenile European sea bass (*Dicentrarchus labrax*) exposed to contaminated sediments. *Ecotoxicology and Environmental Safety*, 97: 114-123.

[104] Cotou, E., Henry, M., Zeri, C., Rigos, G., Torreblanca, A. and Catsiki, V-A. (2012). Short-term exposure of the European sea bass *Dicentrarchus labrax* to copper-based antifouling treated nets: Copper bioavailability and biomarkers responses. *Chemosphere*, 89: 1091-1097.

[105] Costa, P. M., Neuparth, T. S., Caeiro, S., Lobo, J., Martins, M. and Ferreira, A. M. (2011). Assessment of the genotoxic potential of contaminated estuarine sediments in fish peripheral blood: Laboratory versus *in situ* studies. *Environmental Research*, 111: 25-36.

[106] Benedetti, M., Ciaprini, F., Piva, F., Onorati, F., Fattorini, D. and Notti, A. (2012). A multidisciplinary weight of evidence approach for classifying polluted sediments: Integrating sediment chemistry, bioavailability, biomarkers responses and bioassays. *Environment International*, 38: 17-28.

[107] Schmidt, V., Zander, S., Korting, W., Broeg, K., von Westernhagen, H. and Dizer, H. (2003). Parasites of flounder (*Platichthys flesus* L.) from the German Bight, North Sea, and their potential use in biological effects monitoring. Pollution effects on the parasite community and a comparison to biomarker responses. *Helgoland Marine Research*, 57: 262-271.

[108] Cao, L., Huang, W., Liu, J., Yin, X. and Dou, S. (2010). Accumulation and oxidative stress biomarkers in Japanese flounder larvae and juveniles under chronic cadmium exposure. *Comparative Biochemistry and Physiology Part C*, 151: 386-392.

[109] Piva, F., Ciaprini, F., Onorati, F., Benedetti, M., Fattorini, D. and Ausili, A. (2011). Assessing sediment hazard through a weight of evidence approach with bioindicator organisms: a practical model to elaborate data from sediment chemistry, bioavailability, biomarkers and ecotoxicological bioassays. *Chemosphere*, 83: 475-485.

[110] Vethaak, A. D., Jol, J. G., Meijboom, A., Eggens, M. L., Rheinallt, T. and Wester, P. W. (1996). Skin and liver diseases induced in flounder (*Platichthys flesus*) after long-term exposure to contaminated sediments in largescale mesocosms. *Environmental Health Perspectives*, 104: 1218-1229.

[111] Ribecco, C., Baker, M. E., Sasik, R., Zuo, Y., Hardiman, G. and Carnevali, O. (2011). Biological effects of marine contaminated sediments on *Sparus aurata* juveniles. *Aquatic Toxicology*, 104: 308-316.

[112] Gravato, C., Guimaraes, L., Santos, J., Faria, M., Alves, A. and Guilhermino, L. (2010). Comparative study about the effects of pollution on glass and yellow eels (*Anguilla anguilla*) from the estuaries of Minho, Lima and Douro Rivers (NW Portugal). *Ecotoxicology and Environmental Safety*, 73: 524-533.

[113] Oliva, M., Antonio Perales, J., Gravato, C., Guilhermino, L. and Dolores Galindo-Riano, M. (2012). Biomarkers responses in muscle of Senegal sole (*Solea senegalensis*) from a heavy metals and PAHs polluted estuary. *Marine Pollution Bulletin*, 64: 2097-2108.

[114] Siscar, R., Torreblanca, A., Palanques, A. and Sole, M. (2013). Metal concentrations and detoxification mechanisms in *Solea solea* and *Solea senegalensis* from NW Mediterranean fishing grounds. *Marine Pollution Bulletin* 77: 90-99.

[115] De Domenico, E., Mauceri, A., Giordano, D., Maisano, M., Giannetto, A. and Parrino, V. (2013). Biological responses of juvenile European sea bass (*Dicentrarchus labrax*) exposed to contaminated sediments. *Ecotoxicology and Environmental Safety,* 97: 114-123.

[116] Fasulo, S., Mauceri, A., Maisano, M., Giannetto, A., Parrino, V. and Gennuso, F. (2010). Immunohistochemical and molecular biomarkers in *Coris julis* exposed to environmental contaminants. *Ecotoxicology and Environmental Safety,* 73: 873-882.

[117] Kime, D. E. (1995). The effects of pollution on reproduction in fish. *Reviews in Fish Biology and Fisheries*, 5: 52-95.

[118] Hanson, N., Halling, M. and Norin, H. (2013). *Biomarkers for Environmental Monitoring: Suggestions for Norwegian monitoring programs.* Trondheim, Norway: Miljodirectoratet.

[119] de Souza Machado, A. A., Mueller Hoff, M. L., Klein, R. D., Cardozo, J. G., Giacomin, M. M. and Ledes Pinho, G. L. (2013). Biomarkers of waterborne copper exposure in the guppy *Poecilia vivipara* acclimated to salt water. *Aquatic Toxicology*, 138: 60-69.

[120] Downs, C. A., Richmond, R. H., Mendiola, W. J., Rougee, L. and Ostrander G. K. (2006). Cellular physiological effects of the MV Kyowa Violet fuel-oil spill on the hard coral, *Porites lobata*. *Environmental Toxicology and Chemistry*, 25: 3171–3180.

[121] Ahmad, I., Pacheco, M., Santos, M. A. (2006). *Anguilla anguilla* L. oxidative stress biomarkers: an in situ study of freshwater wetland ecosystem (Pateira de Fermentelos, Portugal). *Chemosphere*, 65: 952–962.

[122] Patil, V. K. and David, M. Oxidative stress in freshwater fish, *Labeo rohita* as a biomarker of malathion exposure. (2013). *Environmental Monitoring Assessment*, 184:10191–10199.

[123] Pereira, S., Pinto, A. L., Cortes, R., Fontaínhas-Fernandes, A. and Coimbra, A. M. (2013). Gill histopathological and oxidative stress

evaluation in native fish captured in Portuguese northwestern rivers. *Ecotoxicological Environment Safety*, 90: 157–166.

[124] Williams, C. R. and Gallagher, E. P. (2013). Effects of cadmium on olfactory mediated behaviors and molecular biomarkers in coho salmon (*Oncorhynchus kisutch*). *Aquatic Toxicology*, 141(C): 295–302.

[125] Ringwood, A. H., Hoguet, J., Keppler, C. and Gielazyn, M. (2004). Linkages between cellular biomarker responses and reproductive success in oysters *Crassostrea virginica*. *Marine Environment Research*, 58: 151–155.

[126] Edge, K. J., Johnston, E. L., Roach, A. C. and Ringwood, A. H. (2012). Indicators of environmental stress: cellular biomarkers and reproductive responses in the Sydney rock oyster (*Saccostrea glomerata*). *Ecotoxicology*, 21: 1415–1425.

[127] Blazer, V. S., Iwanowicz, L. R., Henderson, H., Mazik, P. M., Jenkins, J. A., Alvarez, D. A. and Young J. A. (2012). Reproductive endocrine disruption in smallmouth bass (*Micropterus dolomieu*) in the Potomac River basin: spatial and temporal comparisons of biological effects. *Environmental Monitoring Assessment*, 184: 4309–4334.

[128] Balk, L., Hylland, K., Hansson, T., Berntssen, M. H. G, Beyer, J., Jonsson, G., Melbye, A., Grung, M., Torstensen, B. E., Borseth, J. F., Skarphedinsdottir, H. and Klungsoyr, J. (2011). Biomarkers in natural fish populations indicate adverse biological effects in offshore oil production. *PLoS One*, 6: e19735.

[129] Connon, R. E., D'Abronzo, L. S., Hotstetter, N. J., Javidmehr, A., Roby, D. D., Evans, A. F., Loge, F. J. and Werner, I. (2012). Transcription profiling in environmental diagnostics: health assessments in Columbia River basin steelhead (*Oncorhynchus mykiss*). *Environmental Scientific Technology*, 46: 6081–6087.

[130] Connon, R. E., D'Abronzo, L. S., Hotstetter, N. J., Javidmehr, A., Roby, D. D., Evans, A. F., Loge, F. J. and Werner, I. (2012). Transcription profiling in environmental diagnostics: health assessments in Columbia River basin steelhead (*Oncorhynchus mykiss*). *Environmental Scientific Technology*, 46: 6081–6087.

[131] Winston, G. W. (1991). Oxidants and antioxidants in aquatic animals. *Comparative Biochemistry and Physiology Part C*, 100: 173–176.

[132] Kelly, K. A., Havrilla, C. M., Brady, T. C., Abramo, K. H. and Levin, E. D. (1998). Oxidative stress in toxicology: established mammalian and emerging piscine model systems. *Environmental Health Perspectives*, 106: 375–384.

[133] Livingstone, D. R. (2001). Contaminant-stimulated reactive oxygen species production and oxidative damage in aquatic organisms. *Marine Pollution Bulletin*, 42(8): 656–666.

[134] Pandey, S., Parvez, S., Sayeed, I., Hague, R., Bin-Hafeez, B. and Raisuddin, S. (2003). Biomarkers of oxidative stress: a comparative study of river Yamuna fish *Wallago attu* (Bl. & Schn.). *Science of the Total Environment*, 309: 105–115.

[135] Banni, M., Jebali, J., Daubeze, M., Clerandau, C., Guerbej, H., Narbonne, J. F. and Bousetta, H. (2005). Monitoring pollution in Tunisian coasts: application of a classification scale based on biochemical markers. *Biomarkers*, 10: 105–116.

[136] Farombi, E. O., Adelowo, O. A. and Ajimoko, Y. R. (2007). Biomarkers of oxidative stress and heavy metal levels as indicators of environmental pollution in African catfish (*Clarias gariepinus*) from Nigeria Ogun River. *International Journal of Environmental Research and Public Health*, 4: 158–165.

[137] Taylor, A. M. and Maher, W. A. (2010). Establishing metal exposure-dose-response relationships in marine organisms: Illustrated with a case study of cadmium toxicity in *Tellina deltoidalis*. In: Martorino, L.; Puopolo, K., editors. *New Oceanography Research Developments; Marine Chemistry, Ocean Floor Analyses, and Marine Phytoplankton*; Nova Science Publishers; Hauppauge, New York, USA. p. 1-57.

[138] Bacanskas, L. R., Whitaker, J. and Di Guilio, R. T. (2004). Oxidative stress in two populations of killifish (*Fundulus heteroclitus*) with differing contaminant exposure histories. *Marine Environment Research*, 58: 597–601.

[139] Sakuragui, M. M., Paulino, M. G., Pereira, C. D., Carvalho, C. S., Sadauskas-Henrique, H. and Fernandes, M. N. (2013). Integrated use of antioxidant enzymes and oxidative damage in two fish species to assess pollution in man-made hydroelectric reservoirs. *Environment Pollution*, 178: 41–51.

[140] Otto, D. M. and Moon, T. W. (1996). Phase I and II enzymes and antioxidant responses in different tissues of brown bullheads from relatively polluted and non-polluted systems. *Archives of Environmental Contamination and Toxicology.* 31: 141–147.

[141] Marigomez, I., Zorita, I., Izagirre, U., Ortiz-Zarragoitia, M., Navarro, P., Etxebarria, N., Orbea, A., Soto, M. and Cajaraville, M. P. (2013). Combined use of native and caged mussels to assess biological effects of pollution through the integrative biomarker approach. *Aquatic Toxicology*, 13: 32–48.

[142] Moore, M. N., Viarengo, A. G., Somerfield, P. J. and Sforzini, S. (2013). Linking lysosomal biomarkers and ecotoxicological effects at higher biological levels. In: Amiard-Triquet, C.; Amiard, JC.; Rainbow, PS., editors. *Ecological Biomarkers: Indicators of Ecotoxicological Effects.* CRC Press; Boca Raton, FL, USA, p. 107-130.

[143] Domouhtsidou, G. P., Dailianis, S., Kaloyianni, M. and Dimitriadis, V. K. (2004). Lysosomal membrane stability and metallothionein content in *Mytilus galloprovincialis* (L.), as biomarkers: Combination with trace metal concentrations. *Marine Pollution Bulletin*. 48: 572–586.

[144] Frenzilli/Domouhtsidou, G. P., Dailianis, S., Kaloyianni, M. and Dimitriadis, V. K. Lysosomal membrane stability and metallothionein content in *Mytilus galloprovincialis* (L.), as biomarkers: Combination with trace metal concentrations. *Marine Pollution Bulletin*. 48: 572–586.

[145] Steinert, S. A., Streib-Montee, R., Leather, J. M. and Chadwick, D. B. (1998). DNA damage in mussels at sites in San Diego Bay. *Mutation Research*, 399: 65–85.

[146] Aas, E., Beyer, J., Jonsson, G., Reichert, W. L. and Andersen, O. K. (2001). Evidence of uptake, biotransformation and DNA binding of polyaromatic hydrocarbons in Atlantic cod and corking wrasse caught in the vicinity of an aluminium works. *Marine Environment Research*, 52: 213–229.

[147] Harvey, J. S., Lyons, B. P., Page, T. S., Stewart, C. and Parry, J. M. (1999) An assessment of the genotoxic impact of the Sea Empress oil spill by the measurement of DNA adduct levels in selected invertebrate and vertebrate species. *Mutation Research*, 441: 103–114.

[148] Hook, S. E. (2010). Promise and progress in environmental genomics: a status report on the applications of microarray studies in ecologically relevant fish species. *Journal of Fish Biology*, 77: 1999–2022.

[149] Hu, J. S., Durst, M., Kerb, R., Truong, V., Ma, J. T., Khurgin, E., Balaban, D., Gingera, T. R. and Hoffman, B. B. (2000). Analysis of drug pharmacology towards predicting drug behavior by expression profiling using high-density oligonucleotide arrays. *Annals of the New York Academy of Sciences*, 919: 9–15.

[150] Martin, R., Rose, D., Yu, K. and Barros, S. (2006). Toxicogenomics strategies for predicting drug toxicity. *Pharmacogenomics*, 7: 1003–1016.

[151] Tilton, F. A., Tilton, S. C., Bammler, T. K., Beyer, R. P., Stapleton, P. L., Scholz, N. L. and Gallagher, E. P. Transcriptional impact of organophosphate and metal mixtures on olfaction: copper dominates the chlorpyrifos-induced response in adult zebrafish. *Aquatic Toxicology*, 102: 205–215.

[152] Osborn, H. L. and Hook, S. E. (2013). Using transcriptomic profiles in the diatom *Phaeodactylum tricornutum* to identify and prioritize stressors. *Aquatic Toxicology*, 138: 12–25.

[153] Hook, S. E., Osborn, H. L., Spadaro, D. A. and Simpson, S. L. (2014). Assessing mechanisms of toxicant response in the amphipod *Melita plumulosa* through transcriptomic profiling. *Aquatic Toxicology*, 146: 247–257.

[154] Poynton, H. C., Varshavsky, J. R., Chang, B., Cavigiolio, G., Chan, S., Holman, P. S., Loguinov, A. V., Bauer, D. J., Komachi, K., Theil, E. C., Perkins, E. J., Hughes, O. and Vulpe, C. D. (2007). *Daphnia magna* ecotoxicogenomics provides mechanistic insights into metal toxicity. *Environmental Science and Technology*, 41: 1044–1050.

[155] Poyton, H. C. and Vulpe, C. D. (2009). Ecotoxicogenomics: Emerging technologies for emerging contaminants. *Journal of American Water Resources Association*, 45: 83–96.

[156] Van Straalen, N. M. and Feder, M. E. (2012). Ecological and evolutionary functional genomics – how can it contribute to the risk assessment of chemicals. *Environmental Science and Technology*, 46: 3–9.

[157] Qian, X., Ba, Y., Zhuang, Q. and Zhong, G. (2014). RNA-Seq Technology and Its Application in Fish Transcriptomics. *OMICS: Journal of Integrated Biology*, 18: 98-110.

[158] Iavicoli, I., Leso, V. and Schulte, P. A. (2009). Biomarkers of susceptibility: State of the art and implications for occupational exposure to engineered nanomaterials. *Toxicology and Applied Pharmacology*, 299: 112-124.

[159] Hunter, D. J. (2005). Gene–environment interactions in human diseases. *Nature Reviews Genetics*, 6: 287–298.

[160] Forbes, V. E. and Forbes, T. L. (1997). *Ecotoxicologie, Théorie et Applications.* Paris: Institut National de la Recherche Agronomique. [*Ecotoxicology, Theory and Applications*. Paris: National Institute of Agronomic Research].

[161] Belfiore, N. M., and Anderson, S. L. (1998). Genetic patterns as a tool for monitoring and assessment of environmental impacts: The example of genetic ecotoxicology. *Environmental Monitoring Assessment*, 51: 465–79.

[162] Bagley, M. J. (2002). *Genetic Diversity as an Indicator of Ecosystem Condition and Sustainability: Utility for Regional Assessments of Stream Condition in the Eastern United States*. Cincinnati, OH: US Environmental Protection Agency.

[163] Hauser, L. (2003). Molecular/cellular processes and the population genetics of a species. In: *Effects of Pollution on Fish, Molecular Effects and Population Responses,* ed. A. J. Lawrence and K. L. Hemingway, 256–88. Oxford, UK: Blackwell Science.

[164] Groenendijk, D. (1999). Fluctuating life-history parameters indicating temporal variability in metal adaptation in riverine chironimids. *Archives of Environmental Contamination and Toxicology*, 37: 175–81.

[165] Theodorakis, C. W. (2001). Integration of genotoxic and population genetic endpoints in biomonitoring and risk assessment. *Ecotoxicology*, 10: 245–56.

[166] Nevo, E. (2001). Evolution of genome–phenome diversity under environmental stress. *Proceedings of the National Academy of Sciences of the United States of America National Academy of Sciences*, 98: 6233–6240.

[167] Bakiu, R. (2016). *Molecular Evolution of Pyroglutamylated Famide Peptide and Orphan G Protein Coupled Receptor,* Grin Publisher, Germany.

[168] Van Straalen, N. and Timmermans, M. (2002). Genetic variation in toxicant-stressed populations: An evaluation of the "genetic erosion" hypothesis. *Human and Ecological Risk Assessment*, 8: 983–1002.

[169] Woods, R. E. and Hoffman, A. A. (2000). Evolution in toxic environments: Quantitative versus major gene approaches. In: *Demography in Ecotoxicology*, ed. J. Kammenga and R. Laskowski. Chichester, UK: Wiley.

[170] Schluter, D. (2000). *The Ecology of Adaptative Radiation*. New York: Oxford University Press.

[171] Benton, M. J., and Guttman, S. I. (1992). Allozyme genotype and differential resistance to mercury pollution in the caddisfly, *Nectopsyche albida*: II. Mutilocus genotypes. *Canadian Journal of Fisheries and Aquatic Sciences*, 49: 147–9.

[172] Peles, J. D., Towler, W. I. and Guttman, S. I. (2003). Population genetic structure of earthworms (*Lumbricus rubellus*) in soils contaminated by heavy metals. *Ecotoxicology*, 12: 379–386.

[173] Hahn, M. E. (1998). Mechanisms of innate and acquired resistance to dioxin-like compounds. *Reviews in Toxicology*, 2: 395–443.

[174] Hulla, J. (1999). Symposium overview: The role of genetic polymorphism and repair deficiencies in environmental disease. *Toxicological Science*, 47: 135–43.

[175] Burnett, K. G. (2007). *Fundulus* as the premier teleost model in environmental biology: Opportunities for new insights using genomics. *Comparative Biochemistry and Physiology*, 2(D): 257–86.

[176] Luikart, G. (2003). The power and promise of population genomics: From genotyping to genome typing. *Nature Reviews Genetics*, 4: 981–994.

[177] Skøt, L. (2002). Molecular genecology of temperature response in *Lolium perenne*: 2. Association of AFLP markers with ecogeography. *Molecular Ecology*, 11: 1865–1876.

[178] Wang, Z. (2003). Reconciling actual and inferred population histories in the house finch (*Carpodacus mexicanus*) by AFLP analysis. *Evolution,* 57: 2852–1864.

[179] Williams, L. M., and Oleksiak, M. J. (2008). Signatures of selection in natural populations adapted to chronic pollution. *BMC Evolution and Genetics*, 8: 282–293.

[180] Jaccoud, D. (2001). Diversity arrays: A solid state technology for sequence information independent genotyping. *Nucleic Acids Research*, 29: e25.

[181] Bozinovic, G. and Oleksiak, M. J. (2011). Genomic approaches with natural fish populations from polluted environments. *Environmental Toxicology and Chemistry*, 30: 283–289.

[182] Luikart, G. (2003). The power and promise of population genomics: From genotyping to genome typing. *Nature Reviews Genetics*, 4: 981–994.

[183] Moen, T. (2008). Identification and characterization of novel SNP markers in Atlantic cod: Evidence for directional selection. *BMC Genetics*, 9: 18.

[184] Bickham, J. W. (2011). The four cornerstones of evolutionary toxicology. *Ecotoxicology,* 20: 497–502.

[185] Dubrova, Y. E. (1996). Human minisatellite mutation rate after the Chernobyl accident. *Nature*, 380: 683–686.

[186] Ellegren, H. (1997). Fitness loss and germline mutations in barn swallows breeding in Chernobyl. *Nature*, 389: 593–596.

[187] Yauk, C. L. and Quinn, J. S. (1996). Multilocus DNA fingerprinting reveals high rates of heritable genetic mutations in herring gulls in an industrialized urban site. *Proceedings of the National Academy of Sciences of the United States of America National Academy of Sciences*, 93: 12137–12141.

[188] Yauk, C. L. (2000). Induced minisatellite germline mutations in herring gulls (*Larus argenteus*) living near steel mills. *Mutation Research*, 452: 211–218.

[189] Somers, C. M. (2002). Air pollution induces heritable DNA mutations. *Proceedings of the National Academy of Sciences of the United States of America National Academy of Sciences*, 99: 15904–15907.

[190] Demarais, B. D., Dowling, T. E. and Monckley W. L. (1993). Post-perturbation genetic changes in populations of endangered Virgin River chubs. *Conservation Biology*, 7: 334–41.

[191] Facemire, C. F., Gross, T. S. and Guillette Jr, L. J. (1995). Reproductive impairment in the Florida panther: Nature or nurture? *Environmental Health Perspectives*, 103: 79–86.

[192] Matson, C. W. (2006). Evolutionary toxicology and population genetic effects of chronic contaminant exposure on marsh frogs (*Rana ridibunda*) in Sumgayit, Azerbaijan. *Environmental Health Perspectives,* 114: 547–552.

[193] Taylor, M. and Feyereisen, R. (1996). Molecular biology and evolution of resistance to toxicants. *Molecular Biology and Evolution*, 13: 719–734.

[194] Hebert, P. D. N., and Luiker, M. (1996). Genetic effects of contaminant exposure—Towards an assessment of impacts on animal populations. *Science of the Total Environment*, 191: 23–58.

[195] Bard, S. M. (2000) Multixenobiotic resistance as a cellular defense mechanism in aquatic organisms. *Aquatic Toxicology*, 48: 357–389.

[232] Bebianno, M. J. and Langston, W. J. (1998). Cadmium and metallothionein turnover in different tissues of the gastropod. *Littorina Littorea Talanta*, 46(II): 301–313.

[233] Lionetto, M. G., Giordano, M. E., Caricato, R., Pascariello, M. F., Marinosci, L. and Schettino, T. (2001). Biomonitoring of heavy contamination along the Salento coast (Italy) by metallothionein evaluation in *Mytilus galloprovincialis* and *Mullus barbatus*. *Aquatic Conservation: Marine and Freshwater Ecosystems*, 11: 305–310.

[234] Petrovic, S., Ozretic, B., Krajnovic-Ozretic, M. and Bobinac, D. Lysosomal membrane stability and metallothionein in digestive gland of mussels (*Mytilus galloprovincialis* Lam.) as biomarkers in a field study. *Marine Pollution Bulletin*, 42(XII): 1373–1378.

[235] Rodriguez-Ortega, M. J., Alhama, J., Funes, V., Romero-Ruiz, A., Rodriguez-Ariza, A. and Lopez-Barea, J. (2002). Biochemical biomarkers of pollution in the clam *Chamaclea gallina* from south-Spanish littoral. *Environmental Toxicology and Chemistry*, 21: 542–549.

[236] Ross, K., Cooper, N., Bidwell, J. R. and Elder, J. (2002). Genetic diversity and metal tolerance of two marine species: a comparison between populations from contaminated and reference sites. *Marine Pollution Bulletin*, 44: 671–679.

[237] Carajaville, M. P., Bebianno, M. J., Blasco, J., Porte, C., Sarasquete, C. and Viarengo, A. (2000). The use of biomarkers to assess the impact of pollution in coastal environment of the Iberian Peninsula: a practical approach. *Science of Total Environnment*, 247: 295–311.

[238] Blaise, C., Gagne, F., Pellerin, J., Hansen, P. D. and Trottier, S. (2002). Molluscan shellfish biomarker study of the Quebec, Canada, Saguenau Fjord with the soft-shell clam *Mya arenaria*. *Environmental Toxicology*, 17: 170–186.

[239] Gagne, F., Blaise, C., Aoyama, I., Luo, R., Gagnon, C., Couillard, Y., Campbell, P. G. C. and Salazar, M. (2002). Biomarker study of a municipal effluent dispersion plume in two species of freshwater mussels. *Environmental Toxicology,* 17: 149–159.

[240] Chevre, N., Gagne, F., Gagnon, P. and Blaise, C. (2003). Application of rough sites analysis to identify polluted aquatic sites based on a battery of biomarkers: a comparison with classical methods. *Chemosphere*, 51: 13–23.

[241] Giguere, A., Guillard, Y., Cambell, P. G. C., Perceval, O., Hare, L., Alloul B. P. and Pellerin, J. (2003). Steady-state distribution of metals among metallothionein and other cytosolic ligands and links to cytotoxicity in bivalves living along a polymetallic gradient. *Aquatic Toxicology*, 64: 185–200.

[242] Domouthsidou, G. P., Dailianis, S., Kaloyianni, M. and Dimitriadis, V. K. (2004). Lysosomal membrane stability and metallothionein content in *Mytilus galloprovincialis* (L.) as biomarkers. Combination with trace metals concentrations. *Marine Pollution Bulletin*. 2004, 48: 572–586.

[243] George, S. G., Hodgson, P. A., Tytler, P. and Todd, K. (1996). Inducibility of metallothionein mRNA expression and cadmium tolerance in larvae of marine teleost, the turbot (*Scophthalmus maximus*). *Fundamental and Applied Toxicology*, 33: 91-99.

[244] Smet, H. D., Watcher, B. D., Lobinski, R. and Blust, R. (2001). Dynamics of (Cd, Zn)-metallothioneins in gills, liver and kidney of common carp *Cyprinus carpio* during cadmium exposure. *Aquatic Toxicology*, 52: 269-281.

[245] Choi, C. Y., An, K. W., Nelson, E. R., Habibi, H. R. (2007). Cadmium affects the expression of metallothionein (MT) and glutathione peroxidase (GPX) mRNA in goldfish, *Carassiuss auratus*. *Comparative Biochemistry and Physiology*, 145(C): 595-600.

[246] Rovira, M. S., Fernandez-Diaz, C., Canavate, J. P. and Blasco, J. (2005). Effects on metallothionein levels and other stress defences in *Senegal sole* larvae exposed to cadmium. *Bulletin of Environmental Contamination and Toxicology*, 74: 597-603.

[247] Chen, W. Y., John, J. A. C., Lin, C. H. and Chang, C. Y. (2007). Expression pattern of metallothionein, MTF-1 nuclear translocation, and its DNA binding activity in zebrafish (*Danio rerio*) induced by

zinc and cadmium. *Environmental Toxicology and Chemistry*, 26: 110-117.

[248] Lange, A., Ausseil, O. and Segner, H. (2002). Alterations of tissue glutathione levels and metallothionein mRNA in rainbow trout during single and combined exposure to cadmium and zinc. *Comparative Biochemistry Physiology Part C*, 131: 231-243.

[249] Wu, S. M., Weng, C. F., Yu, M. J., Lin, C. C., Chen, S. T., Hwang, J. C., Hwang, P. P. (1999). Cadmium-inducible metallothionein in tilapia (*Oreochromis mossambicus*). *Bulletin of Environmental Contamination and Toxicology*, 62: 758-768.

[250] Ikebuchi, H., Teshima, R., Suzuki, K., Terao, T. and Yamane, Y. (1986). Simultaneous induction of Pb metallothionein-like protein and Zn-thionein in the liver of rats given lead acetate. *The Biochemical Journal*. 233: 541–546.

[251] Carginale, V., Scudiero, R., Capasso, C., Capasso, A., Kille, P. and di Prisco, G. (1998). Cadmium-induced differential accumulation of metallothionein isoforms in the Antarctic icefish, which exhibits no basal metallothionein protein but high endogenous mRNA levels. *The Biochemical Journal*, 332(II): 475–481.

[252] Riggio, M., Filosa, S., Parisi, E. and Scudiero, R (2003). Changes in zinc, copper and metallothionein contents during oocyte growth and early development of the teleost *Danio rerio* (zebrafish). *Comparative Biochemistry and Physiology C Toxicology and Pharmacology*, 135(II): 191–196.

[253] Eroglu, K., Atli, G., Canli, M. (2005). Effects of metal (Cd, Cu, Zn) interactions on the profiles of metallothionein-like proteins in the Nile fish *Oreochromis niloticus*. *Bulletin of Environmental Contamination and Toxicology*, 75(2): 390–399.

[254] Alvarado, N. E., Quesada, I., Hylland, K., Marigómez, I. and Soto, M. (2006). Quantitative changes in metallothionein expression in target cell-types in the gills of turbot (*Scophthalmus maximus*) exposed to Cd, Cu, Zn and after a depuration treatment. *Aquatic Toxicology*, 77(I): 64–77.

[255] Yudkovski, Y., Rogowska-Wrzesinska, A., Yankelevich, I., Shefer, E., Herut, B. and Tom, M. (2008). Quantitative immunochemical evaluation of fish metallothionein upon exposure to cadmium. *Marine Environmental Research*. 65(V): 427–436.

[256] Langston, W. J., Chesman, B. S., Burt, G. R., Pope, N. D. and Mc Evoy., J. (2002). Metallothionein in liver of eels *Anguilla anguilla* from the Thames Estuary: An indicator of environmental quality? *Marine Environmental Research*, 53(III): 263–293.

[257] Sarkar, A., Ray, D., Shrivastava, A. N. and Sarker, S. (2006). Molecular biomarkers: Their significance and application in marine pollution monitoring. *Ecotoxicology*, 15(IV): 333–340.

[258] Nesto, N., Romano, S., Moschino, V., Mauri, M. and Da Ros, L. (2007). Bioaccumulation and biomarker responses of trace metals and micro-organic pollutants in mussels and fish from the Lagoon of Venice, Italy. *Marine Pollution Bulletin*. 55(X–XII): 469–484.

[259] Fernandes, D., Bebianno, M. J. and Porte, C. (2008). Hepatic levels of metal and metallothioneins in two commercial fish species of the Northern Iberian shelf. *The Science of the Total Environment*, 391(I): 159–167.

[260] Serrano, R., Barreda, M. and Blanes, M. A. (2008). Investigating the presence of organochlorine pesticides and polychlorinated biphenyls in wild and farmed gilthead sea bream (*Sparus aurata*) from the Western Mediterranean sea. *Marine Pollution Bulletin*, 56(V): 963–972.

[261] Minghetti, M., Leaver, M. J., Carpenè, E. and George, S. G. (2008). Copper transporter 1, metallothionein and glutathione reductase genes are differentially expressed in tissues of sea bream (*Sparus aurata*) after exposure to dietary or waterborne copper. *Comparative Biochemistry and Physiology C Toxicology and Pharmacology*, 147(IV): 450–459.

[262] Tom, M., Moran, O., Jabukov, E., Cavari, B. and Rinkevitch, B. (1998). Molecular characterization of metallothionein-cDNA of *Sparus aurata* used for detecting heavy metal pollution along the

Mediterranean coast of Israel. *Marine Pollution Bulletin*, 36: 131–137.

[263] Man, A. K. Y. and Woo, N. Y. S (2008). Upregulation of metallothionein and glucose-6-phosphate dehydrogenase expression in silver sea bream, *Sparus sarba* exposed to sublethal leels of cadmium. *Aquatic Toxicology*, 89: 214-221.

[264] Schlenk, D. and Rice, C. D. (1998). Effect of zinc and cadmium treatment on hydrogen peroxide-induced mortality and expression of glutathione and metallothionein in a teleost hepatoma cell line. *Aquatic Toxicology*, 43: 121-129.

[265] Ghedira, J., Jebali, J., Bouraoui, Z., Banni, M., Guerbej, H. and Boussetta, H. (2010). Metallothionein and metal levels in liver, gills and kidney of *Sparus aurata* exposed to sublethal doses of cadmium and copper. *Fish Physiology and Biochemistry*, 36: 101-107.

[266] Man, A. K. Y. and Woo, N. Y. S. (2008). Upregulation of metallothionein and glucose-6-phosphate dehydrogenase expression in silver sea bream, *Sparus sarba* exposed to sublethal leels of cadmium. *Aquatic Toxicology*, 89: 214-221.

[267] Ghedira, J., Jebali, J., Bouraoui, Z., Banni, M., Guerbej, H. and Boussetta, H. (2011). Metallothionein and metal levels in liver, gills and kidney of *Sparus aurata* exposed to sublethal doses of cadmium and copper. *Fish Physiology and Biochemistry*, 36: 101-107.

[268] Viarengo, A., Ponzano, E., Dondero, F. and Fabbri, R. (1997). A simple spectrophotometric method for metallothionein evaluation in marine organism: an application to Mediterranean and Antarctic mollusks. *Marine Environment Reseacj*, 44: 69–84.

[269] Filipovic, V. and Raspor, B. (2003) Metallothioneinn and metal levels in cytosol of liver, kidney and brain in relation to growth parameters of *Mullus surmuletus* and *Liza aurata* from the eastern Adriatic Sea. *Water Research*, 37: 3253–3262.

[270] Cretì, P., Trinchella, F. and Scudiero, R. (2010). Heavy metal bioaccumulation and metallothionein content in tissues of the sea bream *Sparus aurata* from three different fish farming systems. *Environmental Monitoring and Assessment*, 165: 321–329.

[271] Canli, M. and Atli, G. (2003). The relationships between heavy metal (Cd, Cr, Cu, Fe, Pb, Zn) levels and the size of six Mediterranean fish species. *Environmental Pollution*, 121(I): 129–136.

[272] Usero, J., Izquierdo, C., Morillo, J. and Gracia, I. (2004). Heavy metals in fish (*Solea vulgaris, Anguilla anguilla* and *Liza aurata*) from salt marshes on the southern Atlantic coast of Spain. *Environment International*, 29(VII): 949–956.

[273] Linde, A. R., Sanchez-Galan, S. and Garcia-Vazquez, E. (2004). Heavy metal contamination of European eel (*Anguilla anguilla*) and brown trout (*Salmo trutta*) caught in wild ecosystems in Spain. *Journal of Food Protection*, 467(X): 2332–2336.

[274] Ureña, R., Peri, S., del Ramo, J. and Torreblanca, A. (2007). Metal and metallothionein content in tissues from wild and farmed *Anguilla anguilla* at commercial size. *Environment International*, 33: 532–539.

[275] Ciardullo, S., Aureli, F., Coni, E., Guandalini, E., Iosi, F. and Raggi, A. (2008) Bioaccumulation potential of dietary arsenic, cadmium, lead, mercury, and selenium in organs and tissues of rainbow trout (*Oncorhyncus mykiss*) as a function of fish growth. *Journal of Agricultural and Food Chemistry*, 56(VII): 2442–2451.

[276] Riggio, M., Trinchella, F., Filosa, S., Parisi, E. and Scudiero, R. (2003). Accumulation of zinc, copper, and metallothionein mRNA in lizard ovary proceeds without a concomitant increase in metallothionein content. *Molecular Reproduction and Development*, 66(IV): 374–382.

[277] Trinchella, F., Riggio, M., Filosa, S., Volpe, M. G., Parisi, E. and Scudiero, R. (2006). Cadmium distribution and metallothionein expression in lizard tissues following acute and chronic cadmium intoxication. *Comparative Biochemistry and Physiology C Toxicology and Pharmacology*, 144(III): 272–278.

[278] Filipovic, V. M. and Biserka, R. (2007). Metal exposure assessment in native fish, *Mullus barbatus* L., from the Eastern Adriatic Sea. *Toxicology Letters*, 168: 292–301.

[279] Amérand, A., Vettier, A., Sébert, P. and Moisan, C. (2006). A comparative study of reactive oxygen species in red muscle: Pressure effects. *Undersea and Hyperbaric Medicine*, 33(III): 161–167.

[280] Ghoshal, K. and Jacob, S. T. (2001). Regulation of metallothionein gene expression. *Progress in Nucleic Acid Research and Molecular Biology*, 66: 357–384.

[281] Bakiu, R., Santovito, G., Hoda, A., Shehu, J., Durmishaj, S., Irato, P. and Piccinni, E. (2013). Metallothionein (MT): a good biomarker in marine sentinel species like sea bream (*Sparus aurata*). *Albanian Journal of Agricultural Sciences*, 12(II): 247-253, 2013.

[282] Serén, N., Glaberman, S., Carretero, M. A. (2014). Molecular evolution and functional divergence of the metallothionein gene family in vertebrates. *Journal of Molecular Evolution*, 78: 217.

[283] Vasak, M. and Meloni, G. (2011). Chemistry and biology of mammalian metallothioneins. *Journal of Biological Inorganic Chemistry*, 16: 1067–1078.

[284] Guirola, M., Perez-Rafael, S., Capdevila, M., Palacios, O. and Atrian, S. (2012). Metal dealing at the origin of the Chordata phylum: the metallothionein system and metal overload response in Amphioxus. *PLoS ONE*, 7(VIII): e43299.

[285] Nam, D-H., Kim, E-Y., Iwata, H and Tanabe, S. (2007). Molecular characterization of two metallothionein isoforms in avian species: evolutionary history, tissue distribution profile, and expression associated with metal accumulation. *Comparative Biochemistry and Physiology*, 145: 295–305.

[286] Andreani, G., Santoro, M., Cottignoli, S., Fabbri, M., Carpene, E. and Isani, G. (2007). Metal distribution and metallothionein in loggerhead (*Caretta caretta*) and green (*Chelonia mydas*) sea turtles. *The Science of Total Environment*, 390: 287–294.

[287] Kim, M., Park, K., Park, J. Y. and Kwak, I-S. (2013). Heavy metal contamination and metallothionein mRNA in blood and feathers of black-tailed gulls (*Larus crassirostris*) from South Korea. *Environmental Monitoring Assessment,* 185: 2221–2230.

[288] Riggio, M., Trinchella, F., Filosa, S., Parisi, E. and Scudiero, R. (2003). Accumulation of zinc, copper, and metallothionein mRNA in lizard ovary proceeds without a concomitant increase in metallothionein content. *Molecular Reproduction and Development*, 66: 374–382.

[289] Goldfarb, S. B., Kashlan, O. B., Watkins, J. N., Suaud, L., Yan, W., Kleyman, T. R. and Ronald, C. (2006). Differential effects of Hsc70 and Hsp70 on the intracellular trafficking and functional expression of epithelial sodium channels. *Proceedings of National Academy of Sciences USA*, 103: 5817–5822.

[290] Polier, S., Dragovic, Z., Hartl, F. U. and Bracher, A. (2008). Structural basis for the cooperation of HSP 70 and HSP 110 chaperones in protein folding. *Cell*, 133: 1068–1079.

[291] Liberek, K., Lewandowska, A. and Zietkiewicz, S. (2008). Chaperones in control of protein disaggregation. *EMBO Journal*, 27: 328–335.

[292] Daugaard, M., Rohde, M. and Jäättelä, M. (2007). The heat shock protein 70 family: highly homologous proteins with overlapping and distinct functions. *FEBS Letters*, 581: 3702–3710.

[293] Kiang, J. G. and Tsokos, G. C. (1998). Heat shock protein 70-kDa: molecular biology, biochemistry, and physiology. *Pharmacological Therapy*, 80: 183–201.

[294] Robert, J. (2003). Evolution of heat shock protein and immunity. *Development and Comparative Immunology*, 27: 449–464.

[295] Srivastava, P. (2002). Roles of heat-shock proteins in innate and adaptive immunity. *Nature – Review in Immunology*, 2: 185–194.

[296] Calderwood, S. K., Gong, J., Theriault, J. R., Mamadula, S. S., Gray, P. J. (2008). Cell stress proteins: novel immunotherapeutics. In: Foundation, N. (Ed. 2008), *The Biology of Extracellular Molecular Chaperones*, Chichester. John Wiley & Sons Ltd, New York, 115–131.

[297] Alam, M. U., Harken, J. A., Knorn, A. M., Elford, A. R., Wigmore, K., Ohashi, P. S. and Millar, D. G. (2009). Transgenic expression of

Hsc70 in pancreatic islets enhances autoimmune diabetes in response to beta cell damage. *Journal of Immunology*, 183: 5728–5737.

[298] Milani, V., Noessner, E., Ghose, S., Kuppner, M., Ahrens, B., Scharner, A., Gastpar, R., Issels, R. D. (2002). Heat shock protein 70: role in antigen presentation and immune stimulation. *International Journal of Hyperthermia*, 18: 563–575.

[299] Calderwood, S. K., Theriault, J. R. and Gong, J. (2005). Message in a bottle: role of the 70-kDa heat shock protein family in anti-tumor immunity. *European Journal of Immunology*, 35: 2518–2527.

[300] Radons, J. and Multhoff, G. (2005). Immunostimulatory functions of membrane-bound and exported heat hock protein 70. *Exercise Immunology Review*, 11: 17–33.

[301] Johnson, J. D. and Fleshner, M. (2006). Releasing signals, secretory pathways, and immune function of endogenous extracellular heat shock protein 72. *Journal of Leukocite Biology*, 79: 425–434.

[302] Vega, V. L., Rodriguez-Silva, M., Frey, T., Gehrmann, M., Diaz, J. C., Steinem, C., Multhoff, G., Arispe, N. and De Maio, A. (2008). Hsp70 translocates into the plasma membrane after stress and is released into the extracellular environment in a membrane associated form that activates macrophages. *Journal of Immunology*, 180: 4299–4307.

[303] Elsner, L., Muppala, V., Gehrmann, M., Lozano, J., Malzahn, D., Bickeboller, H., Brunner, E., Zientkowska, M., Herrmann, T., Walter, L., Alves, F., Multhoff, G. and Dressel, R. (2007). The heat shock protein HSP70 promotes mouse NK cell activity against tumors that express inducible NKG2D ligands. *Journal of Immunology*, 179: 5523–5533.

[304] Asea, A., Kraeft, S-K., Kurt-Jones, E. A., Stevenson, M. A., Chen, L. B., Finberg, R. W. and Calderwood, S. K. (2000). Hsp70 stimulates cytokine production through a CD14-dependent pathway, demonstrating its dual role as a chaperone and cytokine. *Nature Medicine*, 6: 435–442.

[305] Mukhopadhyay, I., Nazir, A., Saxena, D. K. and Chowdhuri, D. K. (2003) Heat shock response: hsp70 in environmental monitoring. *Journal of Biochemistry and Molecular Toxicology*, 17: 249–254.

[306] Deane, E. E. and Woo N. Y. S. (2005). Cloning and characterization of the hsp70 multigene family from silver sea bream: modulated gene expression between warm and cold temperature acclimation. *Biochemical and Biophysical Research Communication*, 330: 776–783.

[307] Zafarullah, M., Wisniewski, J., Shworak, N. W., Schieman, S., Misra, S. and Gedamu, L. (1992). Molecular cloning and characterization of a constitutively expressed heat-shockcognate hsc71 gene from rainbow trout. *European Journal of Biochemistry*, 204: 893–900.

[308] Ming, J. H., Xie, J., Xu, P., Liu, W. B., Ge, X. P., Liu, B., Cheng He, Y., Zhou, Y., Pan, Q. L. (2009). Molecular cloning and expression of two HSP70 genes in the Wuchang bream (*Megalobrama amblvcenhala* Yih). *Fish Shellfish Immunology*, 28: 407–418.

[309] Beg, M. U., Al-Subiai, S., Beg, K. R., Butt, S. A., AlJandal, N., AI-Hasan, E. and AI-Hussaini, M. (2010). Seasonal effect on heat shock proteins in fish from Kuwait Bay. *Bulletin of Environment Contamination Toxicology*, 84: 91–95.

[310] Heredia-Middleton, P., Brunelli, J., Drew, R. E. and Thorgaard, G. H. (2008). Heat shock protein (HSP70) RNA expression differs among rainbow trout (*Oncorhynchus mykiss*) clonal lines. *Comparative Biochemistry and Physiology B Biochemistry and Molecular Biology*, 149: 552–556.

[311] Webb, D. and Gagnon, M. M. (2009). The value of stress protein 70 as an environmental biomarker of fish health under field conditions. *Environmental Toxicology*, 24: 287–295.

[312] Giusi, G., Alo, R., Crudo, M., Facciolo, R. M. and Canonaco, M. (2008). Specific cerebral heat shock proteins and histamine receptor cross-talking mechanisms promote distinct lead dependent neurotic responses in teleosts. *Toxicology and Applied Pharmacology*, 227: 248–256.

[313] Dang, W., Hu, Y. H., Zhang, M., Sun, L. (2010). Identification and molecular analysis of a stress inducible Hsp70 from *Sciaenops ocellatus*. *Fish and Shellfish Immunology*, 29 (IV): 600–607.

[314] Eder, K. J., Leutenegger, C. M., Köhler, H. R. and Werner, I. (2009). Effects of neurotoxic insecticides on heat-shock proteins and cytokine transcription in Chinook salmon (*Oncorhynchus tshawytscha*). *Ecotoxicology and Environment. Safeness*, 72: 182–190.

[315] Elibol-Flemming, B., Waldbieser, G. C., Wolters, W. R., Boyle, C. R., Hauson, L. A. (2009). Expression analysis of selected immune-relevant genes in channel catfish during *Edwardsiella ictaluri* infection. *Journal of Aquatic Animals Health*, 21: 23–35.

[316] Kominek, J., Marszalek, J., Neuve, C., Craig, E. A., Barry, L. W. The Complex Evolutionary Dynamics of Hsp70s: A Genomic and Functional Perspective. *Genome Biology and Evolution*, 5(12): 2460–2477.

[317] Fridovich, I. (1995). Superoxide radical and superoxide dismutases. *Annual Reviews on Biochemistry*, 64:97–112.

[318] Keele, B. B. Jr McCord, J. M. and Fridovich. I. (1971). Further characterization of bovine superoxide dismutase and its isolation from bovine heart. *Journal of Biological Chemistry*, 246: 2875–2880.

[319] McCord, J. M. and Fridovich, I. (1969). Superoxide dismutase: anenzymic function for erythrocuprein (*Hemo cuprein*). *Journal of Biological Chemistry* 244: 6049–6055.

[320] Marklund, S. L. (1982). Human copper-containing superoxide dismutase of high molecular weight. *Proceedings of the National Academy of Sciences of the United States of America National Academy of Sciences* 79: 7634–7638.

[321] Bannister, J. V., Bannister, W. H. and Rotilio, G. (1987) Aspects of the structure, function and applications of superoxide disinutase. *Critical Reviews in Biochemistry and Molecular Biology*, 22: 111–180.

[322] James, E. R. (1994). Superoxide dismutase. *Parasitology Today*, 10: 481–484.

[323] Fréalle, E., Noël, C., Nolard, N., Symoens, F., Felipe, M. S., Dei-Cas, E., Camus, D., Viscogliosi, E. and Delhaes, L. (2006) Manganese superoxide dismutase based phylogeny of pathogenic fungi. *Molecular Phylogenetic Evolution*, 41: 28–39.

[324] Santos, R., Bocquet, S., Puppo, A. and Touati, D. (1999). Characterization of an atypical superoxide dismutase from *Sinorhizo biummeliloti*. *Journal of Bacteriology*, 181: 4509–4516.

[325] Sugio, S., Hiraoka, B. Y. and Yamakura, F. (2000). Crystal structure of cambialistic superoxide dismutase from *Porphyromonas gingivalis*. *Europian Journal of Biochemistry*, 267: 3487–3495.

[326] Jackson, S. M. and Cooper, J. B. (1998) An analysis of structural similarity in the iron and manganese superoxide dismutases based on known structures and sequences. *Biometals,* 11: 159–173.

[327] Lah, M. S., Dixon, M. M., Pattridge, K. A., Stallings, W. C., Fee, J. A. and Ludwig, M. L. (1995). Structure-function in *Escherichia coli* iron superoxide dismutase: comparisons with the manganese enzyme from *Thermus thermophilus*. *Biochemistry,* 34: 1646–1660.

[328] Parker, M. W., Blake, C. C., Barra, D., Bossa, F., Schinina, M. E., Bannister, W. H. and Bannister, J. V. (1987). Structural identity between the iron and manganese containing superoxide dismutases. *Protein Engeenering,* 1: 393–397.

[329] Smith, M. W. and Doolittle, R. F. (1992). A comparison of evolutionary rates of the two major kinds of superoxide dismutase. *Journal of Molecular Evolution*, 34: 175–184.

[330] Tainer, J. A., Getzoff, E. D., Richardson, J. S. and Richardson, D. C. (1983) Structure and mechanism of copper, zinc superoxide dismutase. *Nature* 306: 284–287.

[331] Youn, H. D., Kim, E. J., Roe, J. H., Hah, Y. C., Kang, S. O. (1996). A novel nickel-containing superoxide dismutase from *Streptomyces* spp. *Biochemistry Journal*, 318: 889–896.

[332] Harris, E. D. (2000). Cellular copper transport and metabolism. *Annual Review of Nutrition*, 20: 291–310.

[333] Pe˜na, M. M., Lee, J. and Thiele, D. J. (1999). A delicate balance: homeostatic control of copper uptake and distribution. *Journal of Nutrition*, 129: 1251–1260.

[334] Esterbauer, H., Gebicki, J., Puhl, H. and Jürgens, G. (1992). The role of lipid peroxidation and antioxidants in oxidative modification of LDL. *Free Radicals in Biology and Medicine,* 13: 341–390.

[335] Ziouzenkova, O., Sevanian, A., Abuja, P. M., Ramos, P., Esterbauer, H. (1998). Copper can promote oxidation of LDL by markedly different mechanisms. *Free Radicals in Biology and Medicine,* 24: 607–623.

[336] Ferro, D., Bakiu, R., De Pittà, C., Boldrin, F., Cattalini, F., Pucciarelli, S., Miceli, C. and Santovito, G. (2015). Cu,Zn superoxide dismutases from *Tetrahymena thermophila*: molecular evolution and gene expression of the first line of antioxidant defenses. *Protist*, 166(I): 131-145.

[337] Swart, E. C., Bracht, J. R., Magrini, V., Minx, P., Chen, X., Zhou, Y., Khurana, J. S., Goldman, A. D., Nowacki, M., Schotanus, K., Jung, S., Fulton, R. S., Ly, A., McGrath, S., Haub, K., Wiggins, J. L., Storton, D., Matese, J. C., Parsons, L., Chang, W. J., Bowen, M. S., Stover, N. A., Jones, T. A., Eddy, S. R., Herrick, G. A., Doak, T. G., Wilson, R. K., Mardis, E. R. and Landweber, L. F. (2013). The *Oxytricha trifallax* macronuclear genome: a complex eukaryotic genome with 16,000 tiny chromosomes. *PLoS Biology,* 11: e1001473.

[338] Pucciarelli, S., La Terza, A., Ballarini, P., Barchetta, S., Yu, T., Marziale, F., Passini, V., Methé, B., Detrich, H. W. and Miceli, C. (2009). Molecular cold-adaptation of protein function and gene regulation: the case for comparative genomicanalyses in marine ciliated protozoa. *Marine Genomics*, 2: 57–66.

[339] Valbonesi, A. and Luporini, P. (1993). Biology of *Euplotes focardii* an Antarctic ciliate. *Polar Biology*, 13: 489–493.

[340] Lee, Y. M., Friedman, D. J. and Ayala, F. J. (1985). Superoxide dismutase: an evolutionary puzzle. *Proceedings of the National Academy of Sciences of the United States of America*, 82: 824–828.

[341] Chen, C., Gentles, A. J., Jurka, J. and Karlin, S. (2002) Genes, pseudogenes and Alu sequence organization across human chromosomes 21 and 22. *Proceedings of the National Academy of Sciences of the United States of America,* 99: 2930–2935.

[342] Gonc¸alves, I., Duret, L. and Mouchiroud, D. (2000). Nature and struc-ture of human genes that generate retropseudogenes. *Genome Research*, 10: 672–678.

[343] Cherian, M. G., Jayasurya, A. and Bay, B. H. (2003). Metallothioneins inhuman tumors and potential roles in carcinogenesis. *Mutation Research*, 533: 201–209.

[344] Boldrin, F., Santovito, G., Negrisolo, E. and Piccinni, E. (2003). Cloning and sequencing of four new metallothionein genes from *Tetrahymena thermophila* and *T. pigmentosa*: evolutionary relationships in *Tetrahymen*a MT family. *Protist,* 154: 431–442.

[345] Boldrin, F., Santovito, G., Gaertig, J., Wloga, D., Cassidy-Hanley, D., Clark, T. G. and Piccinni, E. (2006). Metallothionein gene from *Tetrahymena thermophila* with a copper-inducible-repressible promoter. *Eukaryotic Cell,* 5: 422–425.

[346] Diaz, S, Amaro, F., Rico, D., Campos, V., Benitez, L., Martin-Gonzales, A., Hamilton, E. P., Orias, E. and Gutierrez, J. C. (2007). *Tetrahymena metallothioneins* fall into two discrete subfamilies. *PLoS ONE,* 2: e291.

[347] Piccinni, E., Bertaggia, D., Santovito, G., Miceli, C., Kraev, A. (1999). *Cadmium metallothionein* gene of *Tetrahymena pyriformis*. *Gene*, 234: 51–59.

[348] Santovito, G., Irato, P., Palermo, S., Boldrin, F., Sack, R., Hunziker, P., Piccinni, E. (2001). Identification, cloning and characterization of a novel copper- metallothionein in *Tetrahymena pigmentosa*. Sequencing of cDNA and expression. *Protist*, 152: 203–213.

[349] Santovito, G., Formigari, A., Boldrin, F., Piccinni, E. (2007). Molecular and functional evolution of *Tetrahymena metallothioneins*: new insights into the gene family of *Tetrahymena thermophila*, 144: 391–397.

[350] Santovito, G., Formigari, A., Boldrin, F. and Piccinni, E. (2007). Molecular and functional evolution of *Tetrahymena metallothioneins*: new insights into the gene family of *Tetrahymena thermophila*. *Comparative Biochemistry and Physiology C,* 144: 397–407.

[351] Shang, Y., Song, X., Bowen, J., Corstanje, R., Gao, Y., Gaertig, J. and Gorovsky, M. A. (2002). A robust inducible-repressible promoter greatly facilitates gene knockouts, conditional expression, and overexpression of homologous and heterologous genes in *Tetrahymena thermophila*. *Proceedings of the National Academy of Sciences of the United States of America,* 99: 3734–3739.

[352] Mossallam, A. A. A. (2007). Identification and assignment of human superoxide dismutase-1 retro pseudo genes using comparative mapping and bioinformatic analysis. *Biotechnology*, 6: 393–407.

[353] Santovito, G., Formigari, A., Boldrin, F., Piccinni, E. (2007). Molecular and functional evolution of *Tetrahymena metallothioneins*: new insights into the gene family of *Tetrahymena thermophila*. *Comparative Biochemistry and Physiology C,* 144: 391–397.

[354] Bordo, D., Djinovic, K. and Bolognesi, M. (1994). Conserved patterns in the Cu, Zn superoxide dismutase family. *Journal of Molecular Evolution,* 238: 366–386.

[355] Fink, R. C. and Scandalios, J. G. (2002). Molecular evolution and structure-function relationships of the superoxide dismutase gene families in angiosperms and their relationship to the reukaryotic and prokaryotic superoxide dismutases. *Archives of Biochemistry and Biophysics*, 399: 19–36.

[356] Zhang, N. N., He, Y. X., Li, W. F., Teng, Y. B., Yu, J., Chen, Y., Zhou, C. Z. (2010). Crystal structures of holo and Cu-deficient Cu/Zn-SOD from the silkworm *Bombyx mori* and the implications inamyotrophic lateral sclerosis. *Proteins,* 78: 1999–2004.

[357] Acworth, I. N., McCabe, D. R. and Maher, T. J. (1997). The Analysis of Free Radicals, their Reaction Products, and Antioxidants. In Baskin SI, Salem H (eds) *Oxidants, Antioxidants, and Free Radicals*. Taylor and Francis, Washington DC, pp 23–77.

[358] Kensler, T. W., Wakabayashi, N. and Biswal, S. (2007). Cell survival responses to environmental stresses via the Keap1–Nrf2–ARE pathway. *Annual Review of Pharmacology and Toxicology*, 47: 89–116.

[359] Yoo, H. Y., Chang, M. S. and Rho, H. M. (1999). Heavy metalmediated activation of the rat Cu/Zn superoxide dismutase gene *via* ametal responsive element. *Molecular Genetics and Genomics* 262: 310–313.

[360] Boldrin, F., Santovito, G., Irato, P. and Piccinni, E. (2002) Metal interaction and regulation of *Tetrahymena pigmentosa* metallothionein genes. *Protist*, 153: 283–291.

[361] Boldrin, F., Santovito, G., Formigari, A., Bisharyan, Y., Cassidy-Hanley, D., Clark, T. G., Piccinni, E. (2008). MTT2, a copper-inducible metallothionein gene from *Tetrahymena thermophila*. *Comparative Biochemistry and Physiology C*, 147: 232–240.

[362] Olszewska, M., Bujarski, J. J. and Kurpisz, M. (2012). P-bodies and their functions during mRNA cell cycle: mini-review. *Cell Biochemistry and Function,* 30: 177–182.

[363] Lavut, A. and Raveh, D. (2012). Sequestration of highly expressed mRNAs in cytoplasmic granules, P-bodies, and stress granules enhances cell viability. *PLoS Genetics,* 8: e1002527.

[364] Sunkar, R., Kapoor, A. and Zhu, J. K. (2006). Post-transcriptional induction of two Cu/Zn superoxide dismutase genes in *Arabidopsisis* mediated by downregulation of miR398 and important for oxidative stress tolerance. *Plant Cell,* 18: 2051–2065.

[365] Mikonranta, L., Mappes, J., Kaukoniitty, M. and Freitak, D. (2014). Insect immunity: oral exposure to a bacterial pathogen elicits free radical response and protects from a recurring infection. *Frontiers in Zoology*, 11: 23.

[366] Behrens, S., Peuß, R., Milutinovi, B., Eggert, H., Esser, D. and Rosenstiel, P. (2014). Infection routes matter in population-specific responses of the red flour beetle to the entomopathogen *Bacillus thuringiensis*. *BMC Genomics* 15: 445.

[388] Ferro, K., Ferro, D., Corra, F., Bakiu, R., Santovito, G., Kurtz, J. (2017). Cu,Zn SOD genes in *Tribolium castaneum*: evolution, molecular characterisation and gene expression during immune priming. *Frontiers in Immunology*, 8:1811.

[389] Halliwell, B. and Gutteridge, J. M. C. (1992). Biologically relevant metal ion-dependent hydroxyl radical generation an update. *FEBS Letters* 307: 108–112.

[390] Kamata, H. and Hirata, H. (1999). Redox regulation of cellular signalling. *Cell Signaling*, 11: 1–14.

[391] Kamata, H. and Hirata, H. (1999). Redox regulation of cellular signalling. *Cell Signaling*, 12: 1–14.

[392] Mills, G. C. (1957). Hemoglobin catabolism I. Glutathione peroxidase, an erythrocyte enzyme which protects hemoglobin from oxidative breakdown. *Journal of Biological Chemistry,* 229: 189–197.

[393] Little, C. and O'Brien, P. J. (1968). An intracellular GSH-peroxidase with a lipid peroxide substrate. *Biochemical and Biophysical Research Communications* 31: 145–150.

[394] Christophersen, B. O. (1969). Reduction of X-ray-induced DNA and thymine hydroperoxides by rat liver glutathione peroxidase. *Biochimica et Biophysica Acta* (BBA)—*Nucleic Acids and Protein Synthesis* 186: 387–389.

[395] Toppo, S., Vanin, S., Bosello, V. and Tosatto, V. C. (2008). Evolutionary and structural insights into the multifaceted glutathione peroxidase (GPx) superfamily. *Antioxidants and Redox Signaling*, 10: 1501–1514.

[396] Maiorino, M., Roche, C., Kiess, M., Koenig, K., Gawlik, D., Matthes, M., Naldini, E., Pierce, R. and Flohe, L. (1996). A selenium-containing phospholipid-hydroperoxide glutathione peroxidase in *Schistosoma mansoni*. *European Journal of Biochemistry*, 238: 838–844.

[397] Gladyshev, V. N. (2006). Selenoproteins and selenoproteoms. In Hatfield, D. L., M. J. Berry & V. N. Gladyshev (eds*), Selenium: Its Molecular Biology and Role in Human Health.* Springer, New York: 99–110.

[398] Tosatto, S. C., Bosello, V., Fogolari, F. Mauri, P., Roveri, A., Toppo, A., Flohe, F., Ursini, F. and Maiorino, F. (2008). The catalytic site of glutathione peroxidases. *Antioxidants and Redox Signaling*, 10: 1515–1526.

[399] Margis, R., Dunand, C., Teixeira, F. K. and Margis-Pinheiro, M. (2008). Glutathione peroxidase family—an evolutionary overview. *FEBS Journal* 275: 3959–3970.

[400] Sattin, G., Bakiu, R., Tolomeo, A. M., Carraro, A., Coppola, D., Ferro, D., Patarnello, T. and Santovito, G. (2015). Characterization and expression of a new cytoplasmic glutathione peroxidase 1 gene in the Antarctic fish *Trematomus bernacchii. Hydrobiologia,* 761(1).

[401] Legg, S., Briegleb, B., Chang, Y., Chassignet, E. P., Danabasoglu, G. Ezer, T., Gordon, A. L., Griffies, S., Hallberg, R., Jackson, L., Large, W., Kmen, T. M., Peters, H. J., Price, U., Riemenschneider, W., Wu, Xu, X. and Yang, J. (2009). Improving oceanic overflow representation in climate models: the gravity current entrainment climate process team. *Bulletin of the American Meteorological Society,* 90: 657–670.

[402] Clark, M. S. and Peck, L. S. (2009). Triggers of the HSP70 stress response: environmental responses and laboratory manipulation in an Antarctic marine invertebrate (*Nacella concinna*). *Cell Stress Chaperones*, 14: 649–660.

[403] Dirks, R. C., Faiman, M. D. and Huyser, E. S. (1982). The roles of lipid free radical initiator and oxygen on the kinetics of lipid peroxidation. *Toxicology and Applied Pharmacology* 63: 21–28.

[404] Thompson, J. L., See, V. H., Thomas, P. M. and Schuller, K. A. (2010). Cloning and characterization of two glutathione peroxidase cDNAs from southern bluefin tuna (*Thunnus maccoyii*). *Comparative Biochemistry and Physiology B,* 156: 287–297.

[405] Pacitti, D., Wang, T. Page, M. M., Martin, S. A., Sweetman, J., Feldmann, J. and Secombes, C. J. (2013). Characterization of cytosolic glutathione peroxidase and phospholipid-hydroperoxide glutathione peroxidase genes in rainbow trout (*Oncorhynchus mykiss*)

and their modulation by *in vitro* selenium exposure. *Aquatic Toxicology*, 130–131: 97–111.

[406] Malandrakis, E. E., Exadactylos, A., Dadali, O. Golomazou, E., Klaoudatos, S. and Panagiotaki, P. (2014). Molecular cloning of four glutathione peroxidase (GPx) homologs and expression analysis during stress exposure of the marine teleost *Sparus aurata*. *Comparative Biochemistry and Physiology B*, 168: 53–61.

[407] Allendorf, F. W. and Thorgaard, G. H. (1984). Tetraploidy and evolution of salmonid fishes. In Turner, B. J. (ed.), *Evolutionary Genetics of Fish*. Plenum Press, New York: 1–53.

[408] Stegeman, J. J., Goldstone, J. V. and Hahn, M. E. (2010). Perspectives on zebrafish as a model in environmental toxicology. In: Perry, S. F., Ekker, M., Farrell, A. P. & C. J. Brauner (eds), *Zebrafish, Fish Physiology* 29: 367–439.

[409] Near, T. J., Dornburg, A., Kuhn, K. L., Eastman, J. T., Pennington, J. N. Patarnello, T., Zane, L., Fernandez, D. A. and Jones, C. D. (2012). Ancient climate change, antifreeze, and the evolutionary diversification of Antarctic fishes. *Proceedings of the National Academy of Sciences of the United States of America* 109: 3434–3439.

[410] Thompson, M. R. A. (2004). Geological and palaeoenvironmental history of the Scotia Sea region as a basis for biological interpretation. *Deep-Sea Research,* 51(II): 1467–1487.

[411] Clarke, A., Barnes, D. K. A. and Hodgson, D. A. (2005). How isolated is Antarctica? *Trends in Ecology and Evolution,* 20: 1–3.

[412] Anderson, J. B., Shipp, S. S., Lowe, A. L., Wellner, J. S. and Mosola, A. B. (2002). The Antarctic ice sheet during the last glacial maximum and its subsequent retreat history: a review. *Quarternary Science Reviews* 21: 49–70.

[413] Ansaldo, M., Luquet, C. M., Evelson, P. A., Polo, J. M., Llesuy, S. (2000). Antioxident levels from different Antarctic fish caught around South Georgia Island and Shag Rocks. *Polar Biology,* 23: 160–165.

[414] Santovito, G., Piccinni, E., Boldrin, F. and Irato, P. (2012). Comparative study on metal homeostasis and detoxification in two

Antarctic teleosts. *Comparative Biochemistry and Physiology C*, 155: 580–586.

[415] Enzor, L. A. and Place, S. P. (2014). Is warmer better? Decreased oxidative damage in notothenioid fish after long-term acclimation to multiple stressors. *Journal of Experimental Biology*, 217: 3301–3310.

[416] Eastman, J. T. (1993). *Antarctic Fish Biology. Evolution in a Unique Environment.* Academic Press, New York.

[417] Eastman, J. T. and De Vries, A. L. (1982). Buoyancy studies of notothenioid fishes in McMurdo Sound, Antarctica. *Copeia* 1982: 385–393.

[418] Sidell, B. D. (1991). Physiological roles of high lipid content in tissues of Antarctic fish species. In di Prisco, G., B. Maresca & B. Tota (eds), *Biology of Antarctic Fish*. Springer, Berlin, Heidelberg, New York: 220–231.

[419] Crockett, E. L. and Sidell, B. D. (1993). Substrate selectivities differ for hepatic mitochondrial and peroxisomal b/oxidation in an Antarctic fish, *Notothenia gibberifrons*. *Biochemical Journal,* 289: 427–433.

[420] Roberfroid, M. and Buc Calderon, P. (1995). *Free radicals and oxidation phenomena in biological systems*. Dekker, New York.

[421] Santovito, G., Cassini, A. and Piccinni, E. (2006). Cu, Zn superoxide dismutase from *Trematomus bernacchii*: functional conservation and erratic molecular evolution in Antarctic teleosts. *Comparative Biochemistry and Physiology C* 143: 444–454.

[422] Benedetti, M., Martuccio, G., Fattorini, D., Canapa, A., Barucca, M., Nigro, M. and Regoli, F. (2007). Oxidative and modulatory effects of trace metals on metabolism of polycyclic aromatic hydrocarbons in the Antarctic fish *Trematomus bernacchii*. *Aquatic Toxicology* 85: 167–175.

[423] Olszewska, M., Bujarski, J. J. and Kurpisz, M. (2012). P-bodies and their functions during mRNA cell cycle: mini-review. *Cell Biochemistry and Function*, 30: 177–182.

[424] Lavut, A. and Raveh, D. (2012). Sequestration of highly expressed mRNAs in cytoplasmic granules, P-bodies, and stress granules enhances cell viability. *PLoS Genetics* 8: e1002527.

[425] Gon, O. and Heemstra, P. C. (1990). Fishes of the Southern Ocean. *J. L. B. Smith Institute of Ichthyology*, Grahamstown.

[426] Kim K. W., Kim I. H., Lee K. Y., Rhee S. G. and Stadtman E. R. (1988). The isolation and purification of a specific protector protein which inhibits enzyme inactivation by a thiol/Fe(III)/O_2 mixed-function oxidation system. *Journal of Biological Chemistry*, 263: 4704–4711.

[427] Chae H. Z., Chung S. J. and Rhee S. G. (1994). Thioredox independent peroxide reductase from yeast. *Journal of Biological Chemistry*, 269: 27670-27678.

[428] Wood Z. A., Poole L. B. and Karplus P. A. (2003). Peroxiredoxin evolution and the regulation of hydrogen peroxide signaling. *Science*, 300: 650–653.

[429] Rhee S. G. and Woo H. A. (2011). Multiple functions of peroxiredoxins: peroxidases, sensors and regulators of the intracellular messenger H2O2, and protein chaperones. *Antioxidants and Redox Signaling*, 15: 781–794.

[430] Fisher, A. B. (2011). Peroxiredoxin 6: a bifunctional enzyme with glutathione peroxidase and phospholipase A2 activities. *Antioxidants and Redox Signaling*, 15: 831-844.

[431] Cha, M. K., Yun, C. H. and Kim, I. H. (2000). Interaction of human thiol-specific antioxidant protein 1 with erythrocyte plasma membrane. *Biochemistry*, 39: 6944-6950.

[432] Cao Z., Lindsay J. G. and Isaacs N. W. (2007). Mitochondrial peroxiredoxins. *Subcellular Biochemistry*, 44: 295-315.

[433] Okado-Matsumoto A., Matsumoto A., Fujii J. and Taniguchi N. (2000). Peroxiredoxin IV is a secretable protein with heparin binding properties under reduced conditions. *Journal of Biochemistry*, 127: 493–501.

[434] Rhee S. G., Woo H. A., Kil I. S. and Bae S. H. (2012). Peroxiredoxin functions as a peroxidase and a regulator and sensor of local peroxides. *Journal of Biological Chemistry*, 287: 4403–4410.

[435] Chen J. W., Dodia C., Feinstein S. I., Jain M. K. and Fisher A. B. (2000). 1-Cys peroxiredoxin, a bifunctional enzyme with glutathione

peroxidase and phospholipase A2 activities. *Journal of Biological Chemistry*, 275: 28421-28427.

[436] Jang H. H., Lee K. O., Chi Y. H., Jung B. G., Park S. K., Park J. H., Lee J. R., Lee S. S., Moon J. C., Yun J. W., Choi Y. O., Kim W. Y., Kang J. S., Cheong G-W, Yun D-J, Rhee S. G., Cho M. J. and Lee S. Y. (2004). Two enzymes in one; two yeast peroxiredoxins display oxidative stress-dependent switching from a peroxidase to a molecular chaperone function. *Cell*, 117: 625–635.

[437] Phalen, T. J., Weirather, K., Deming, P. B., Anathy, V., Howe, A. K., van der Vliet A., Jönsson, T. J., Poole, L. B. and Heintz, N. H. (2006). Oxidation state governs structural transitions in peroxiredoxin II that correlate with cell cycle arrest and recovery. *Journal of Cell Biology*, 175: 779–789.

[438] Hall, A., Karplus, P. A. and Poole, L. B. (2009). Typical 2-Cys peroxiredoxins – structures, mechanisms and functions. *FEBS Journal*, 276: 2469–2477.

[439] Hanschmann, E-A, Godoy, J. R., Berndt, C., Hudemann, C. and Lillig, C. H. (2013). Thioredoxins, glutaredoxins, and peroxiredoxins--molecular mechanisms and health significance: from cofactors to antioxidants to redox signaling. *Antioxidants and Redox Signaling*, 19: 1539-1605.

[440] Hamilton, R. T., Walsh, M. E. and Van Remmen, H. (2012). Mouse Models of Oxidative Stress Indicate a Role for Modulating Healthy Aging. *Journal of Clinical and Experimental Pathology*, 4(S): 1–14.

[441] Ishii T., Warabi, E. and Yanagawa, T. (2012). Novel roles of peroxiredoxins in inflammation, cancer and innate immunity. *Journal of Clinical Biochemistry and Nutrition*, 50: 91-105.

[442] Conway Morris, S., Peel, J. S., Higgins, A. K., Soper, N. J. and Davis N. C. (1987). A burgess shale-like fauna from the lower Cambrian of north Greenland. *Nature*, 345: 802-805.

[443] Valentine, J. W. (1997). Cleavage patterns and the topology of metazoan tree of life. *Proceedings of the National Academy of Sciences U. S. A.*, 94: 8001–8005.

[444] Edgar, R. S., Green, E. W., Zhao, Y., van Ooijen, G., Olmedo, M., Qin, X., Xu, Y., Pan, M., Valekunja, U. K., Feeney, K. A., Maywood, E. S., Hastings, M. H., Baliga, N. S., Merrow, M., Millar, A. J., Johnson, C. H., Kyriacou, C. P., O'Neill, J. S. and Reddy, A. B. (2012). Peroxiredoxins are conserved markers of circadian rhythms. *Nature*, 485: 459-464.

[445] Halliwell, B. and Gutteridge, J. M. C. (1999). Free Radicals in Biology and Medicine, 3rd ed. Oxford: Oxford Science Publications. Copley S. D., Novak W. R. P. and Babbitt P. C. 2004. Divergence of Function in the Thioredoxin Fold Suprafamily: Evidence for Evolution of Peroxiredoxins from a Thioredoxin-like Ancestor. *Biochemistry*, 43: 13981-13995.

[446] Knoops, B., Loumaye, E. and Van Der Eecken, V. (2007). Evolution of the peroxiredoxins. *Subcellular Biochemistry*, 44: 27–40.

[447] Pérez-Sánchez, J., Bermejo-Nogales, A., Calduch-Giner, J. A., Kaushik, S. and Sitjà-Bobadilla, A. (2011). Molecular characterization and expression analysis of six peroxiredoxin paralogous genes in gilthead sea bream (*Sparus aurata*): insights from fish exposed to dietary, pathogen and confinement stressors. *Fish Shellfish Immunology,* 31: 294–302.

[448] Bakiu, R. and Santovito, G. (2015). New Insights into the Molecular Evolution of Metazoan Peroxiredoxins. *Acta Zoologica Bulgarica* 67(2).

[449] Saccoccia, F., Di Micco, P., Boumis, G., Brunori, M., Koutris, I., Miele, A. E., Morea, V., Sriratana, P., Williams, D. L., Bellelli, A. and Angelucci, F. (2012). Moonlighting by different stressors: crystal structure of the chaperone species of a 2-Cys peroxiredoxin. *Cell*, 20: 429–439.

[450] Hofmann, B., Hecht, H. J. and Flohé, L. (2002). Peroxiredoxins. *Biological Chemistry*, 383: 347–364.

[451] Kosiol, C., Vinar, T., da Fonseca, R. R., Hubisz, M. J., Bustamante, C. D., Nielsen, R. and Siepel, A. (2008). Patterns of positive selection in six mammalian genomes. *PLoS Genetics*, 4: e1000144.

[452] König, J, Galliardt, H, Jütte, P, Schäper, S, Dittmann, L. and Dietz, K. J. (2013). The conformational bases for the two functionalities of 2-cysteine peroxiredoxins as peroxidase and chaperone. *Journal of Experimental Botany* 64(11): 3483–3497.

[453] Randall, L. M., Ferrer-Sueta, G. and Denicola, A. (2013). Peroxiredoxins as preferential targets in H_2O_2-induced signaling. *Methods in Enzymology,* 527: 41–63.

[454] Tolomeo, A. M., Carraro, A., Bakiu, R., Toppo, S., Place, S. P., Ferro, D. and Santovito, G. (2015). Peroxiredoxin 6 from the Antarctic emerald rockcod: molecular characterization of its response to warming. *Journal of Comparative Physiology B*,186(1).

[455] Liu, N. N., Liu, Z. S., Lu, S. Y., Hu, P., Li, Y. S., Feng, X. L., Zhang, S. Y., Wang, N., Meng, Q. F., Yang, Y. J., Tang, F., Xu, Y. M., Zhang, W. H., Guo, X., Chen, X. F., Zhou, Y. and Ren, H. L. (2015). Full-length cDNA cloning, molecular characterization and differential expression analysis of peroxiredoxin 6 from *Ovis aries*. *Veterinary Immunology and Immunopathology*, 164(IV): 208–219.

[456] Rahaman, H., Zhou, S., Dodia, C., Feinstein, S. I., Huang, S., Speicher, D. and Fisher, A. B. (2012). Increased phospholipase A2 activity with phosphorylation of peroxiredoxin 6 requires a conformational change in the protein. *Biochemistry*, 51(XXVII): 5521–5530.

[457] Manevich, Y., Feinstein, S. I. and Fisher, A. B. (2004). Activation of the antioxidant enzyme 1-CYS peroxiredoxin requires glutathionylation mediated by heterodimerization with pGST. *Proceedings of the National Academy of Sciences of the United States of America*, 101(11): 3780–3785.

[458] Fisher, A. B. (2011). Peroxiredoxin 6: a bifunctional enzyme with glutathione peroxidase and phospholipase A2 activities. *Antioxid Redox Signaling,* 15(III): 831–844.

[459] Manevich, Y., Reddy, K. S., Shuvaeva, T., Feinstein, S. I. and Fisher, A. B. (2007). Structure and phospholipase function of peroxiredoxin 6: identification of the catalytic triad and its role in phospholipid substrate binding. *Journal of Lipid Research*, 48(X): 2306–2318.

[460] Nikapitiya, C., De Zoysa, M., Whang, I., Kim, C. G., Lee Y. H., Kim, S. J. and Lee, J. (2009). Molecular cloning, characterization and expression analysis of peroxiredoxin 6 from disk abalone *Haliotis discus* and the antioxidant activity of its recombinant protein. *Fish Shellfish Immunol*ogy, 27(II): 239–249.

[461] Smeets, A., Loumaye, E., Clippe, A., Rees, J. F., Knoops, B. and Declercq, J. P. (2008). The crystal structure of the C45S mutant of annelid *Arenicola marina* peroxiredoxin 6 supports its assignment to the mechanistically typical 2-Cys subfamily without any formation of toroid-shaped decamers. *Protein Science*, 17(4): 700–710.

[462] David, E., Tanguy, A. and Moraga, D. (2007). Peroxiredoxin 6 gene: a new physiological and genetic indicator of multiple environmental stress response in Pacific oyster *Crassostrea gigas*. *Aquatic Toxicology*, 84(III): 389–398.

[463] Zheng, W. J., Hu, Y. H., Zhang, M. and Sun, L. (2010). Analysis of the expression and antioxidative property of a peroxiredoxin 6 from *Scophthalmus maximus*. *Fish Shellfish Immunol*ogy, 29(2): 305–311.

[464] De Zoysa, M., Ryu, J. H., Chung, H. C., Kim, C. H., Nikapitiya, C., Oh, C., Kim, H., Revathy, K. S., Whang, I. and Lee, J, (2012). Molecular characterization, immune responses and DNA protection activity of rock bream (*Oplegnathus fasciatus*), peroxiredoxin 6 (Prx6). *Fish Shellfish Immunology,* 33(1): 28–35.

[465] Lin, Y., Dobbs, G. H., Devries, A. L. (1974). Oxygen consumption and lipid content in red and white muscles of Antarctic fishes. *Journal of Experimental Zoology,* 189(3): 379–386.

[466] Clarke, A., Doherty, N., De Vries, A. L. and Eastman, J. T. (1984) Lipid content and composition of three species of Antarctic fish in relation to buoyancy. *Polar Biology*, 3: 77–83.

[467] Johnston, I. I., Calvo, J. and Guderley, Y. H. (1998) Latitudinal variation in the abundance and oxidative capacities of muscle mitochondria in perciform fishes. *Journal or Experimental Biology*, 201(I):1–12.

[468] Pörtner, H. O., Langenbuch, M. and Michaelidis, B. (2005). Synergistic effects of temperature extremes, hypoxia and increases in

CO_2 on marine animals: from earth history to global change. *Journal of Geophysics Research*, 110: C09S10.
[469] Roberfroid, M. and Buc Calderon, P. (1995). *Free radicals and oxidation phenomena in biological systems*. Dekker, New York.
[470] Abele, D. and Puntarulo, S. (2004). Formation of reactive species and induction of antioxidant defence systems in polar and temperate marine invertebrates and fish. *Comparative Biochemistry and Physiology*, 138A: 405–415.
[471] Albergoni, V., Cassini, A., Coppellotti, O., Favero, M., Favero, N., Irato, P., Piccinni, E. and Santovito, G. (2000). Physiological responses to heavy metal and adaptation to increase oxygen partial pressure in Antarctic fish and protozoa. *Italian Journal of Zoology*, 67(SI): 1–11.
[472] Santovito, G., Irato, P., Piccinni, E. and Albergoni, V. (2000). Relationship between metallothionein and metal contents in red-blooded and white-blooded Antarctic teleosts. *Polar Biology*, 23(6): 383–391.
[473] Huang, F. and Nau, W. M. (2003). A conformational flexibility scale for amino acids in peptides. *Angewandte Chemie International Edition*, 42(XX): 2269–2272.
[474] Gekko, K., Kamiyama, T., Ohmae, E. and Katayanagi, K. (2000). Single amino acid substitutions in flexible loops can induce large compressibility changes in dihydrofolate reductase. *Journal of Biochemistry*, 128(1): 21–27.
[475] Römisch, K. and Matheson, T. (2003). Cell biology in the Antarctic: studying life in the freezer. *Nature Cell Biology*, 5(1): 3–6.
[476] Siddiqui, K. S. and Cavicchioli, R. (2006). Cold-adapted enzymes. *Annual Review of Biochemistry*, 75: 403–433.
[477] Enzor, L. A., Zippay, M. L. and Place, S. P. (2013). High latitude fish in a high CO_2 world: synergistic effects of elevated temperature and carbon dioxide on the metabolic rates of Antarctic notothenioids. *Comparative Biochemistry and Physiology Part A*, 164(I): 154–161.

[478] Burnett, K. G. (2007). *Fundulus* as the premier teleost model in environmental biology: Opportunities for new insights using genomics. *Comparative Biochemistry and Physiology*, 2(D): 257–86.

[479] Fent, K. and Sumpter, J. P. (2011). Progress and promises in toxicogenomics in aquatic toxicology: Is technical innovation driving scientific innovation? *Aquatic Toxicology*, 105: 25–39.

[480] Nikinmaa, M. and Rytkönen K. T. (2011). Functional genomics in aquatic toxicology—Do not forget the function? *Aquatic Toxicology*, 105: 16–24.

[481] Whitehead, A. (2010). Comparative transcriptomics implicates mechanisms of evolved pollution tolerance in a killifish population. *Molecular Ecology*, 19: 5186–5203.

[482] Smolders, R., Bervoets, L. and Blust, R. (2004). *In situ* and laboratory bioassays to evaluate the impact of effluent discharges on receiving aquatic ecosystems. *Environmental Pollution*, 132: 231–243.

[483] Phillips, D. J. H. (1994). Macrophytes as biomonitors of trace metals. In: *Biomonitoring of Coastal Waters and Estuaries*, ed. K. J. M. Kramer, 85–106. Boca Raton, FL: CRC Press.

[484] Dauvin, J. C., Bellan, D. and Bellan-Santini, D. (2010). Benthic indicators: From subjectivity to objectivity— Where is the line? *Marine Pollution Bull.* 60:947–53.

[485] Chapman, P. M. (1994). Refinery water (intake and effluent) quality: Update of 1970s with 1990s toxicity testing. *Environmental Toxicology and Chemistry*, 13: 897–909.

[486] Michel, X. (1998). Indicateurs biochimiques de pollution des ecosystems côtiers: Experience du groupe interface chimie-biologie des ecosystems marins (GICBEM). In: *Utilisation de biomarqueurs pour la surveillance de la qualité de l'environnement,* ed. L. Lagadic, T. Caquet, J. C. Amiard, and F. Ramade, 9–32. Paris: Lavoisier Tec & Doc [Michel, X. (1998). Biochemical indicators of pollution of coastal ecosystems: Experience of the chemistry-biology interface group of marine ecosystems (GICBEM). In: *Use of biomarkers for monitoring the quality of the environment,* ed. L. Lagadic, T. Caquet, J. C. Amiard, and F. Ramade, 9-32. Paris: Lavoisier Tec & Doc.]

[487] Chapman, P. M. (2002). Integrating toxicology and ecology: Putting the "eco" into ecotoxicology. *Marine Pollution Bulletin*, 44: 7–15.

[488] Mc Carty, L. S., Power, M. and Munkittrick, K. R. (2002). Bioindicators versus biomarkers. In: Ecological Risk Assessment. *Human Ecology Risk. Assessment*, 8: 159–64.

[489] Galloway, T. S. (2004). Ecosystem management bioindicators: The ECOMAN project—A multibiomarker approach to ecosystem management. *Marine Environment Research*, 58: 233–237.

[490] Tabor, G. M. and Aguirre, A. A. (2004). Ecosystem health and sentinel species: Adding an ecological element to the proverbial "canary in the mineshaft." *Eco Health* 1: 226–228.

[491] De Lange, H. J., Sala, S. and Faber, J. H. (2010). Ecological vulnerability in risk assessment—A review and perspectives. *The Science of Total Environment*, 408: 3871–3879.

[492] Luoma, S. N. and Rainbow, P. S. (2010). Linking new science frontiers in metal ecotoxicology to better risk management: Lateral thinking. In: *Surface Chemistry, Bioavailability and Metal Homeostasis in Aquatic Organisms: An Integrated Approach,* ed. N. R. Bury and R. D. Handy, 1–28. *Essential reviews in Experimental Biology*, vol. 2. London, UK: Society for Experimental Biology (SEB).

[493] Chapman, P. M. and Wang, F. (2001). Assessing sediment contamination in estuaries. *Environment Toxicology and Chemistry*, 20: 3–22.

[494] Tixier, G. (2011). Ecological risk assessment of urban stormwater ponds: Literature review and proposal of a new conceptual approach providing ecological quality goals and the associated bioassessment tools. *Ecological indicators*, 11: 1497–1506.

[495] Lam, P. K. and Gray, J. S. (2001). Predicting effects of toxic chemicals in the marine environment. *Marine Pollution Bulletin*, 42: 169–173.

[496] Rand, G. M. (1985). Behavior. In: *Fundamentals of Aquatic Toxicology: Methods and Applications*, ed. G. M. Rand and S. R. Petrocelli, 221–63. Washington, D.C.: Hemisphere Publishing.

[497] NRC, Committee on Animals as Monitors of Environmental Hazards. (1991). *Animals as Sentinels of Environmental Health Hazards.* Washington, D.C.: National Research Council, National Academy Press.

[498] White, G. J. (2004). Selection of ecological indicators for monitoring terrestrial systems. In: *Environmental Monitoring,* ed. G. B. Wiersma, 263–82. Boca Raton, FL: CRC Press.

[499] NOAA. (2005). https://coastalscience.noaa.gov/category/stressor-impacts/ Updated 18 August 2018.

[500] De Kock, W. C. and Kramer, K. J. M. (1994). Active biomonitoring (ABM) by translocation of bivalve Molluscs. In: *Biomonitoring of Coastal Waters and Estuaries,* ed. K. J. M. Kramer, 51–84. Boca Raton, FL: CRC Press.

[501] Olivier, F., Ridd, M. and Klumpp, D. (2002). The use of transplanted cultured tropical oysters (*Saccostrea commercialis*) to monitor Cd levels in North Queensland coastal waters (Australia). *Marine Pollution Bulletin*, 44: 1051–1062.

[502] Wepener, V. (2005). Active biomonitoring in freshwater environments: Early warning signals from biomarkers in assessing biological effects of diffuse sources of pollutants. *Physics and Chemistry of the Earth*, 30: 751–761.

[503] Ji, Y. (2010). Fish transplantation and stress-related biomarkers as useful tools for assessing water quality. *Journal of Environmental Sciences*, 22: 1826–1832.

[504] Lu, G. H. (2010). Active biomonitoring of complex pollution in Taihu Lake with *Carassius auratus*. *Chemosphere*, 79: 588–594.

[505] RINBIO. (2002). www.corse.eaufrance.fr/rlm/reseaux-resultats-surveillance/rinbio.htm.9k; Updated 25/01/2002.

[506] Geffard, A. (2004). Comparative study of metal handling strategies in bivalves *Mytilus edulis* and *Crassostrea gigas*: A multidisciplinary approach. *Journal of the Marine Biological Association of the UK*. 84: 1–10.

[507] Takamura, N., Kasai, F. and Watanabe, M. M. (1989). Effects of Cu, Cd and Zn on photosynthesis of freshwater benthic algae. *Journal of Applied Phycology*, 1: 39–52.

[508] Berthet, B. (1992). Bioaccumulation, toxicity and physico-chemical speciation of silver in bivalves molluscs: Ecotoxicological and health consequences. *The Science of Total Environment*, 125: 97–122.

[509] Wallner-Kersanach, M. (2000). Accumulation and elimination of trace metals in a transplantation experiment with *Crassostrea rhizophorae*. *Archives of Environmental Contamination and Toxicology*, 38: 40–45.

[510] Nigro, M. (2006). Cellular biomarkers for monitoring estuarine environments: Transplanted versus native mussels. *Aquatic Toxicology*, 77: 339–347.

[511] Regoli, F. and Principato, G. (1995). Glutathione, glutathione-dependent and antioxidant enzymes in mussel, *Mytilus galloprovincialis*, exposed to metals under field and laboratory conditions: Implications for the use of biochemical biomarkers. *Aquatic Toxicology*, 31: 143–164.

[512] Bolognesi, C. (2004). Genotoxicity biomarkers in *Mytilus galloprovincialis*: Wild versus caged mussels. *Mutation Research*, 552: 153–162.

[513] Damiens, G. (2006). Metal bioaccumulation and metallothionein concentrations in larvae of *Crassostrea gigas*. *Environmental Pollution*, 140: 492–499.

[514] Collin, H. (2010). Response of the Pacific oyster *Crassostrea gigas*, Thunberg 1793, to pesticide exposure under experimental conditions. *Journal of Experimental Biology*, 213: 4010–4017.

[515] Camus, L. (2004). Total oxyradical scavenging capacity responses in *Mytilus galloprovincialis* transplanted into the Venice lagoon (Italy) to measure the biological impact of anthropogenic activities. *Marine Pollution Bulletin*, 49: 801–808.

[516] Honkoop, P. J. C. (2003). Appropriate experimental design for transplanting mussels (*Mytilus* sp.) in analyses of environmental

stress. *Journal of Experimental Marine Biology and Ecology*, 297: 253–268.

[517] Couillard, Y. (1995). Field transplantation of freshwater bivalve, *Pyganodon grandis*, across a metal contamination gradient: I. Temporal changes in metallothionein and metal (Cd, Cu and Zn) concentrations in soft tissues. *Canadian Journal of Fisheries and Aquatic Sciences*, 52: 690–702.

[518] Geffard, A., Amiard, J. C. and Amiard-Triquet, C. (2002). Kinetics of metal elimination in oysters from a contaminated estuary. *Comparative Biochemistry and Physiology*, 131: 281–93.

[519] Evtushenko, Z. S., Belcheva, N. N. and Lukyanova, O. N. (1986). Cadmium accumulation in organs of the scallop *Mizuhopecten yessoensis*: 2. Subcellular distribution of metals and metal-binding proteins. *Comparative Biochemistry and Physiology*, 83C: 377–383.

[520] Couillard, Y. (1995). Field transplantation of freshwater bivalve, *Pyganodon grandis*, across a metal contamination gradient: II. Metallothionein response to Cd and Zn exposure, evidence for cytotoxicity, and links to effects at higher levels of biological organization. *Canadian Journal of Fisheries and Aquatic Sciences*, 52: 703–715.

[521] Mouneyrac, C., Berthet, B. and Amiard, J. C. (1999). Cd distribution in the tissues of oysters (*Crassostrea gigas*) exposed chronically *in situ*. *Water Air Soil Pollution*, 112: 187–196.

[522] Regoli, F. (2004). Time-course variations of oxyradical metabolism, DNA integrity and lysosomal stability in mussels, *Mytilus galloprovincialis*, during a field translocation experiment. *Aquatic Toxicology*, 68: 167–178.

[523] Camus, L. (2004). Total oxyradical scavenging capacity responses in *Mytilus galloprovincialis* transplanted into the Venice lagoon (Italy) to measure the biological impact of anthropogenic activities. *Marine Pollution Bulletin*, 49: 801–808.

[524] Hagger, J. A. (2009). Application of biomarkers to assess the condition of European Marine Sites. *Environmental Pollution*, 157: 2003–2010.

[525] Bellwood, D. R., Hoey, A. Q.S. and Choat, J. H. (2003). Limited functional redundancy in high diversity systems: Resilience and ecosystem function on coral reefs. *Ecology Letters*, 6: 281–285.

[526] Bielmyer, G. K. (2010). Differential effects of copper on three species of scleractinian corals and their algal symbionts (*Symbiodinium* spp.). *Aquatic Toxicology*, 97: 125–133.

[527] De Pirro, M. and Marshall, D. J. (2005). Phylogenetic differences in cardiac activity, metal accumulation and mortality of limpets exposed to copper: A prosobranch–pulmonate comparison. *Journal of Experimental Marine Biology and Ecology*, 322: 29–37.

[528] Scarfe, A. D. (1982). Locomotor behavior of four marine teleosts in response to sublethal copper exposure. *Aquatic Toxicology*, 2: 335–353.

[529] Kalman, J. (2010). Comparative toxicity of cadmium in the commercial fish species *Sparus aurata* and *Solea senegalensis*. *Ecotoxicology and Environmental Safety*, 73: 306–311.

[530] Eyckmans, M. (2011). Exposure to waterborne copper reveals differences in oxidative stress response in three freshwater fish species. *Aquatic Toxicology*, 103: 112–120.

[531] Keizer, J. (1995). Enzymological differences of AChE and diazinon hepatic metabolism: Correlation of in vitro data with the selective toxicity of diazinon to fish species. *Science of the Total Environment*, 171: 213–20.

[532] Oliveira, M. M. (2007). Brain acetylcholinesterase as a marine pesticide biomarker using Brazilian fishes. *Marine Environmental Research*, 63: 303–312.

[533] Gomez Gesteira, J. L., and Dauvin, J. C. (2000). Amphipods are good bioindicators of the impact of oil spills on soft-bottom macrobenthic communities. *Marine Pollution Bulletin*, 40: 1017–1027.

[534] Wentsel, R., McIntosh, A. and Anderson, V. (1977). Sediment contamination and benthic macroinvertebrate distribution in a metal-impacted lake. *Environmental Pollution*, 14: 187–193.

[535] Dauvin, J. C. and Ruellet, T. (2007). Polychaete/amphipod ratio revisited. *Marine Pollution Bulletin*, 55: 215–224.

[536] Andrade, H. and Renaud, P. E. (2011). Polychaete/amphipod ratio as an indicator of environmental impact related to offshore oil and gas production along the Norwegian continental shelf. *Marine Pollution Bulletin*, 62: 2836–2844.

[537] Pessoa, P. C. (2011). Cholinesterase inhibition and behavioral toxicity of carbofuran on *Oreochromis niloticus* early life stages. *Aquatic Toxicology*, 105: 312–320.

[538] Alzieu, C., Quiniou, F. and Delesmont, R. (2003). L'approche globale: Scores de risque, tests de toxicité, indices biotiques. In: *Bioévaluation de la qualité environnementale des sédiments portuaires et des zones d'immersion*, coord. C. Alzieu, 175–90. Ifremer. [The global approach: Risk scores, toxicity tests, biotic indices. In: *Bioassessment of the environmental quality of port sediments and dumping zones,* coord. C. Alzieu, 175-90. Ifremer.]

[539] Lotufo, G. R. and Fleeger, J. W. (1997). Effects of sediment-associated phenanthrene on survival, development and reproduction of two species of meiobenthic copepods. *Marine Ecology Progress Series*, 15: 91–102.

[540] Stubblefield, W. A. (1999). Acclimation-induced changes in the toxicity of zinc and cadmium to rainbow trout. *Environmental Toxicology and Chemistry*, 18: 2875–2881.

[541] Kammenga, J. E. and Riksen, J. A. G. (1996). Comparing differences in species sensitivity to toxicants: Phenotypic plasticity versus concentration–response relationships. *Environmental Toxicology and Chemistry*, 15: 1649–1653.

[541] Gagnon, C. and Fisher, N. S. (1997). The bioavailability of sediment-bound Cd, Co and Ag to the mussel *Mytilus edulis*. *Canadian Journal of Fisheries and Aquatic Sciences*, 54: 147–156.

[542] Bernes, C. and Naylor, M. (1998). *Persistent Organic Pollutants. A Swedish View of an International Problem.* Stockholm: Swedish Environmental Protection Agency.

[543] Chapman, P. M. and Wang, F. (2001). Assessing sediment contamination in estuaries. *Environmental Toxicology and Chemistry*, 20: 3–22.

[544] Moreira, S. M. (2006). Effects of estuarine sediment contamination on feeding and on key physiological functions of the polychaete *Hediste diversicolor*. Laboratory and *in situ* assays. *Aquatic Toxicology*, 78: 186–201.

[545] Poirier, L. (2006). A suitable model for the biomonitoring of trace metal bioavailabilities in estuarine sediments: The annelid polychaete *Nereis diversicolor*. *Journal of the Marine Biological Association of the United Kingdom*, 86: 71–82.

[546] Coelho, J. P. (2008). The role of two sediment-dwelling invertebrates on the mercury transfer from sediments to estuarine trophic web. *Estuarine, Coastal and Shelf Science*, 78: 505–512.

[547] Solé, M., Kopecka-Pilarczyk, J. and Blasco, J. (2009). Pollution biomarkers in two estuarine invertebrates, *Nereis diversicolor* and *Scobicularia plana*, from a Marsh ecosystem in SW Spain. *Environment International*, 35: 523–531.

[548] Fossi Tankoua, O. (2010). *Towards a comprehensive methodology for assessing the health status of coastal and estuarine ecosystems.* ICES CM 2010/F: 15.

[549] Fossi Tankoua, O. (2012). Intersite variations of a battery of biomarkers at different levels of biological organisation in the estuarine endobenthic worm *Nereis diversicolor* (Polychaeta, Nereididae). *Aquatic Toxicology*, 114-115: 96-103.

[550] Piva, F. (2011). Assessing sediment hazard through a weight evidence approach with bioindicator organisms: A practical model to elaborate data from sediment chemistry, bioavailability, biomarkers and ecotoxicological bioassays. *Chemosphere*, 83: 475–485.

[551] Benedetti, M. (2012). A multidisciplinary weight of evidence approach for classifying polluted sediments: Integrating sediment chemistry, bioavailability, biomarkers responses and bioassays. *Environment International*, 38: 17–28.

[552] Maes, G. E. (2005). The catadromous European eel *Anguilla anguilla* (L.) as a model for freshwater evolutionary ecotoxicology: Relationship between heavy metal bioaccumulation, condition and genetic variability. *Aquatic Toxicology*, 73: 99–114.

[553] Belpaire, C. (2011). What can we learn from monitoring PCBs in the European eel? A Belgian experience. *Environment International*, 37: 354–364.

[554] Valbonesi, P. (2011). Cholinesterase activities and sensitivity to pesticides in different tissues of silver European eel, *Anguilla anguilla*. *Comparative Biochemistry and Physiology*, 154C: 353–359.

[555] Van der Oost, R. (1996). Biomonitoring of aquatic pollution with feral eel (*Anguilla anguilla*): II. Biomarkers: Pollution-induced biochemical responses. *Aquatic Toxicology*, 36: 189–222.

[556] Depledge, M. H. and Hopkin, S. P. (1995). Methods to assess effects on brackish, estuarine, and near coastal water organisms. In: *SCOPE 53. Methods to Assess the Effects of Chemicals on Ecosystems,* ed. R. A. Lithurst, P. Bourdeau, and R. J. Tardiff, 125–49.

[557] UNEP/RAMOGE. (1999). *Manuel sur les biomarqueurs recommandés pour le programme de biosurveillance MED POL.* Athènes: PNUE, 41–92. [*Manual on recommended biomarkers for the MED POL biomonitoring program.* Athens: UNEP, 41-92.].

[558] Roméo, M. (1994). Heavy metal content and biotransformation enzymes in two fish species from the NW Mediterranean. *Marine Ecology Progress Series*, 107: 15–22.

[559] Castro, B. B. (2004). An *in situ* bioassay integrating individual and biochemical responses using small fish species. *Ecotoxicology*, 13: 667–681.

[560] Roméo, M. (2003). Mussel transplantation and biomarkers as useful tools for assessing water quality in the NW Mediterranean. *Environmental Pollution*, 122: 369–378.

[561] Buffet, P. E. (2011). Behavioural and biochemical responses of two marine invertebrates *Scrobicularia plana* and *Hediste diversicolor* to copper oxide nanoparticles. *Chemosphere*, 84: 166–174.

[562] Durou, C. (2007). Biomonitoring in a clean and a multi-contaminated estuary based on biomarkers and chemical analyses in the endobenthic worm *Nereis diversicolor*. *Environmental Pollution*, 148: 445–458.

[563] Chapman, P. M., Farrel, M. A. and Brinkhurst, R. O. (1982). Effects of species interactions on the survival and respiration of *Limnodrilus*

hoffmeisteri and *Tubifex tubifex* (Oligochaeta, tubificidae) exposed to various pollutants and environmental factors. *Water Research*, 16: 1405–1408.

[564] Keilty, T. J., White, D. S. and Landrum P. F. (1988). Sublethal responses to endrin in sediment by *Limnodrilus hoffmeisteri* (Tubificidae), and in mixed-culture with *Stylodrilus heringianus* (Lumbriculidae). *Aquatic Toxicology*, 13: 227–249.

ABOUT THE AUTHOR

Rigers Bakiu, PhD

Department of Aquaculture and Fisheries,
Faculty of Agriculture and Environment,
Agricultural University of Tirana,
Tirane, Albania
Email: rigers.bakiu@ubt.edu.al

Rigers Bakiu is an Associate Professor at the Department of Aquaculture and Fisheries, where he is also the Head of the Aquaculture Research Group from 2014. In 2004, Rigers Bakiu was graduated on Molecular Biology at the University of Padova (Italy), where he also performed his Master of Science studies on Molecular Biology. Furthermore, at the University of Padova, from 2007 to 2010 Rigers Bakiu performed his PhD studies and he obtained the PhD certification on Biosciences. Rigers Bakiu continued with the post-PhD studies at the Department of Biomedical Sciences of University of Padova from 2010 to 2012. He obtained the title of Associate Professor in 2015 and actually he is the lecturer of Aquatic Toxicology and other aquatic biotechnologies modules. In the University, the research activity is focused on Marine Aquaculture, Applied Biotechnologies on Aquaculture and Studies about the Molecular Toxicology on Aquatic

Biological Systems. Rigers Bakiu is the author of several students' text books and he has published more than 40 scientific papers (including impacted factor journals). Furthermore, he has been involved as expert in other national and international expert on Environmental Sciences and Blue Growth Disciplines Projects. On April 2017 Rigers Bakiu won the nation price "Most Successful Young Scientist 2016" given to him by the Albanian National Academy of Sciences and actually he is the Head of the Section about Technical and Natural Sciences of the Albanian Young Academy.

INDEX

C

D

G

H

I

T

U

V

W

Z

Aquatic Ecosystems: Influences, Interactions and Impact on the Environment

Editor: Marvin Snyder

Series: Environmental Health - Physical, Chemical and Biological Factors

Book Description: The seagrass meadows of the Northeast Coast of Brazil have the greatest area, biomass, biometrics and diversity of the Brazilian Coast. In this region, the seagrass diversity and plasticity may vary due to differences in environmental factors.

Hardcover ISBN: 978-1-63484-686-8
Retail Price: $110

Neotropical and Caribbean Aquatic Mammals: Perspectives from Archaeology and Conservation Biology

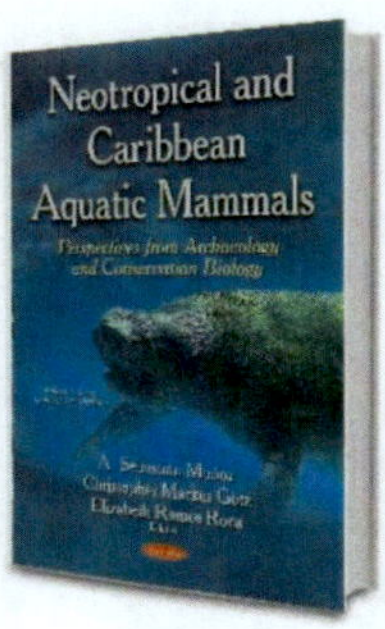

Editors: A. Sebastián Muñoz (CONICET-Universidad Nacional de Córdoba), Christopher Markus Götz (Facultad de Ciencias Antropológicas, Universidad Autónoma de Yucatán), and Elizabeth Ramos Roca (Departamento de Antropología, Universidad de los Andes)

Series: Animal Science, Issues and Research

Book Description: The relationships between humans and aquatic mammals in the Neotropics has been important since archaic times in the American continent, but also varied across time and space, a fact taken on by this publication.

Softcover ISBN: 978-1-63483-929-7
Retail Price: $105

Water Resource and the Aquatic Environment: Current Issues and Options for Sustainable Management

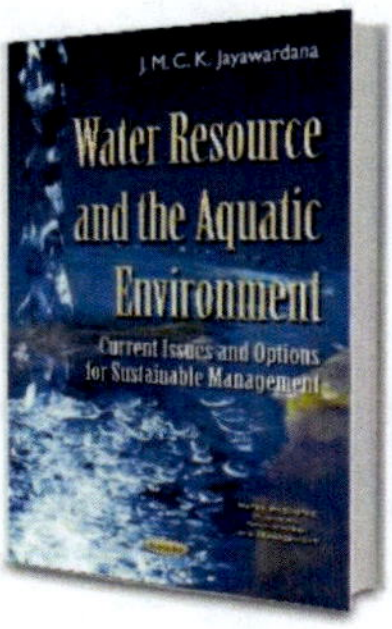

Author: J.M.C.K. Jayawardana (Department of Natural Resources, Faculty of Applied Sciences, Sabaragamuwa University of Sri Lanka, Belihuloya, Sri Lanka)

Series: Water Resource Planning, Development and Management

Book Description: One of the main challenges faced by humans today is finding a balance between economic development of respective countries and sustainable utilization of Earth's resources. This book discusses the current status of water resources on Earth and challenges water resource conservation in the first chapter.

Softcover ISBN: 978-1-63485-881-6
Retail Price: $95

Aquatic Invasive Species: Federal Activities and Costs of Addressing Threats and Impacts

Editor: Ernestine Sandoval

Series: Marine Biology

Book Description: This book examines how much Task Force member agencies expended addressing aquatic invasive species for fiscal years 2012-2014; activities conducted by Task Force member agencies and challenges in addressing aquatic invasive species; and the extent to which the Task Force has measured progress in achieving the goals of its 2013-2017 strategic plan.

Softcover ISBN: 978-1-63485-391-0
Retail Price: $95